Plasma Physics

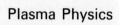

Plasma Physics

R.A. Cairns, BSc, PhD

Senior Lecturer in Applied Mathematics
University of St Andrews

Blackie

Glasgow and London
Distributed in the USA and Canada by
Heyden & Son, Inc.
Philadelphia

Blackie & Son Limited
Bishopbriggs, Glasgow G64 2NZ

Furnival House, 14–18 High Holborn, London WC1V 6BX

Distributed in the USA and Canada by
Heyden & Son, Inc.
247 South 41st St
Philadelphia, PA 19104

First published 1985

British Library Cataloguing in Publication Data

Cairns, R.A.
Plasma physics.
1. Plasma diagnostics 2. Nuclear fusion
I. Title
539.7'64 QC718.5.D5

ISBN 0-216-91779-4
ISBN 0-216-91783-2 Pbk

For the USA and Canada, International Standard Book Numbers are
0-86 344-026-6
0-86 344-027-4 (Pbk)

Photosetting by Thomson Press (India) Limited, New Delhi
Printed in Great Britain by Bell & Bain (Glasgow) Ltd

Contents

Preface

This book is intended as an introduction to plasma physics at a level suitable for advanced undergraduates or beginning postgraduate students in physics, applied mathematics or astrophysics. The main prerequisite is a knowledge of electromagnetism and of the associated mathematics of vector calculus. SI units are used throughout. There is still a tendency amongst some plasma physics researchers to cling to c.g.s. units, but it is the author's view that universal adoption of SI units, which have been the internationally agreed standard since 1960, is to be encouraged.

After a short introductory chapter, the basic properties of a plasma concerning particle orbits, fluid theory, Coulomb collisions and waves are set out in Chapters 2–5, with illustrations drawn from problems in nuclear fusion research and space physics. The emphasis is on the essential physics involved and the theoretical and mathematical approach has been kept as simple and intuitive as possible. An attempt has been made to draw attention to areas of current research and to present plasma physics as a developing subject with many areas of uncertainty, and not as something to be set forth on 'tablets of stone'.

Chapter 6 deals with the theory of nonlinear problems, discussing analytical methods and also the fundamentals of computational techniques, using both fluid and particle codes. In Chapter 7 the principles of some of the diagnostic methods used to obtain information about laboratory and space plasmas are discussed. Finally, Chapter 8 considers some applications of plasma physics to nuclear fusion research, using both magnetic and inertial confinement, and to some problems in space physics.

Most of the chapters end with a few problems, intended to illustrate and test the student's knowledge of the material in them. In an introductory text I have not felt it appropriate to give a comprehensive list of references, but a list of further reading is provided to guide the interested reader either to more advanced texts or to recent review articles.

I should like to extend thanks to many friends, colleagues, teachers and students, particularly at Glasgow University, St Andrews University and Culham Laboratory and amongst the users of the Central Laser Facility at the Rutherford Appleton Laboratory, from whom I have learnt much about plasma physics. Finally, thanks are due to my wife for her help in preparing the manuscript and in eliminating some of the more convoluted examples of English prose from the original draft.

R.A.C.

1 Introduction

1.1 Nature and occurrence of plasmas

A plasma is essentially a gas consisting of charged particles, electrons and ions, rather than neutral atoms or molecules. In general the plasma is electrically neutral overall, but the existence of charged particles means that it can support an electric current and react to electric and magnetic fields. It cannot, however, be treated simply as an ordinary gas which is electrically conducting. There is a very fundamental difference between a neutral gas and a plasma, resulting from the very different nature of the inter-particle forces in the two cases. In the former the forces are very strong, but of short range, so the dynamics of a gas is dominated by two-body, billiard-ball-like collisions. In a plasma the forces are Coulomb forces, which are comparatively weak and of long range. This makes possible a variety of collective effects in a plasma, involving the interaction of a large number of particles, and makes plasma physics a rich and complicated subject.

It is a commonplace to point out in introductions to plasma physics that most of the matter in the universe is in the plasma state. Since, however, we live in one of the small regions of the universe where matter is predominantly solid, liquid or gas, concern with the properties of plasma is quite recent and has mainly been stimulated by its importance in space physics and in the development of controlled nuclear fusion.

Since most matter outside the lower layers of the Earth's atmosphere is ionized, its relevance to the first of these is obvious. In interplanetary space and in the upper layers of the ionosphere ionization is mainly produced by ultra-violet radiation, and in the diffuse plasmas which result the rate of recombination of electrons and ions is low.

Nuclear fusion, on which we shall elaborate in the next section, is the process by which stars generate their energy but which so far has only been exploited by man in an uncontrolled fashion. The main obstacle in the way of harnessing this energy source is the fact that the reactions will take place at a useful rate only if the temperature of the reacting material is of the order of 10^8 K. Material at this temperature is ionized, since the thermal energy is well above that required to strip electrons from atoms, and research has centred on using magnetic fields to confine and control the hot plasma. Early attempts to do this soon revealed that a plasma was a much more subtle and complicated system than had been thought, and triggered off a programme of theoretical and experimental research into its properties which still continues. In space physics, satellites are being used to obtain ever more detailed data about the plasma in the vicinity of the Earth, while our knowledge of more distant regions is also being extended by improved observational techniques. Such observations reveal that space plasmas show just as rich a variety of phenomena as laboratory plasmas.

The aim of this book is to acquaint the reader with some of the most important properties of a plasma and in the course of doing so to point out ways in which these properties are relevant to laboratory or naturally occurring plasmas. Many of the illustrations will be drawn from applications connected with the controlled fusion programme, since this undoubtedly has been and remains the driving force behind much of the research on plasmas.

1.2 Controlled nuclear fusion

Because the most strongly bound nuclei are those in the middle of the periodic table, nuclear energy can be released in two ways. These are fusion of light nuclei into heavier ones, and fission of heavy nuclei, the latter being the process which fuels the current generation of nuclear power stations. To obtain energy in a controlled way from fusion, the most suitable reactions involve fusion of hydrogen isotopes to helium. That which requires the lowest temperature involves deuterium and tritium reacting to form helium and a neutron

$$ {}_1D^2 + {}_1T^3 \rightarrow {}_2He^4 + n + 17.56\,\mathrm{MeV}, $$

and if fusion is eventually demonstrated to be feasible it will probably be with this system. Since tritium is an unstable radioactive gas, deuterium–deuterium fusion

$$ {}_1D^2 + {}_1D^2 \rightarrow {}_2He^4 $$

may be more desirable in the long run, but it requires a higher temperature.

In order to understand the problems which stand in the way of controlled nuclear fusion it is necessary to consider the conditions under which a reaction yielding a net power output can take place. For fusion to occur, the reacting nuclei must have enough energy to overcome their Coulomb repulsion and to approach sufficiently closely for there to be a reasonable probability of fusion. For the D–T reaction the cross-section, measuring the probability of the reaction's taking place, has a maximum when the particle energy, in the rest frame of the centre of mass of the two particles, is of the order of 10 keV. As a result, an effective power output requires that the particle thermal energies be of this order. Generally temperatures are measured in eV, the energy equivalent of the temperature, κT, with κ Boltzmann's constant, being expressed in these units. Since 1 eV is equivalent to about 10^4 K, an effective fusion reactor requires a temperature of about 10^8 K to be sustained.

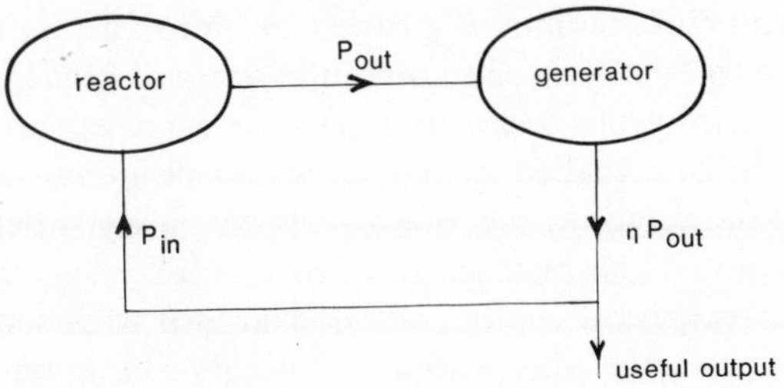

Figure 1.1 Schematic reactor system with energy flow.

Supposing that this can be done we may obtain a further condition to be satisfied if we consider the energy flow in the schematic reactor system shown in Figure 1.1.

In the reactor let us suppose that n is the ion density, there being an equal density of electrons, so that at a temperature T the thermal energy density is $3n\kappa T$. No confinement system is perfect and there is always loss of particles and energy across the magnetic field lines to the walls.

If the energy loss time associated with this is τ, then the rate of loss of thermal energy per unit volume is $3n\kappa T/\tau$. There is also a substantial loss by radiation, which we denote by P_{rad} per unit volume. Plasmas under fusion conditions are almost transparent to radiation, so the radiation loss is a volume rather than a surface effect. Finally we shall suppose that nuclear reactions produce a power P_{nuc} per unit volume.

In order to maintain the temperature, the input power must balance the radiation and diffusion losses, so that, per unit volume

$$P_{in} = 3n\kappa T/\tau + P_{rad}.$$

As well as the nuclear power, the radiated power and that lost by transport across the containing field contribute to the output of the reactor, which is thus

$$P_{out} = 3n\kappa T/\tau + P_{rad} + P_{nuc}.$$

This output, in the form of heat, has to be converted to electricity by a conventional turbine system, with an efficiency η which is typically about 30%, so that the output from this stage is ηP_{out}. Part of this power, P_{in}, must be recirculated within the system to maintain the temperature of the reacting material, so the final useful power output is

$$\begin{aligned} P &= \eta P_{out} - P_{in} \\ &= \eta P_{nuc} - (1-\eta)P_{rad} - (1-\eta)\frac{3n\kappa T}{\tau} \end{aligned} \tag{1.1}$$

If the system is to generate any useful power this must be positive. Now, both P_{nuc} and P_{rad} are proportional to the square of the density, but at temperatures in the range below about 10 keV P_{nuc} increases much more rapidly with temperature than P_{rad} which is proportional to $T^{1/2}$. If the temperature can be kept high enough, of the order of 10 keV, then the combination of the first two terms in (1.1) is positive. Given that this temperature is maintained, the first two terms, which we recall are proportional to n^2, can be made to outweigh the last one if $n\tau$ is big enough. Thus we arrive at the other condition which must be satisfied in addition to the requirement of a high enough temperature. This condition, known as the Lawson criterion, sets a minimum value on the product of density and energy containment time. Its numerical value is around $10^{20}\,\mathrm{m^{-3}\,s}$.

We have discussed this criterion in terms of a steady-state reactor with a magnetic confinement system. However, similar energy balance considerations lead to the same condition being obtained for pulsed systems. One concept which is currently receiving much attention is that of inertial confinement, in which a small target pellet is heated and compressed by a laser or particle beam. No attempt is made to contain the resulting small dense plasma and the confinement time is just the finite time which it takes to fly apart because of its inertia. While magnetic confinement systems aim for densities of the order $10^{20}\,\mathrm{m^{-3}}$ and confinement times of the order of seconds, inertial confinement aims for densities of about a thousand times that of the solid target material and correspondingly short confinement times. Fusion systems will be discussed in more detail in Chapter 8.

1.3 Fluid and kinetic descriptions of a plasma

In order to gain a theoretical understanding of the behaviour of a plasma we must formulate a set of equations to describe it, then try to solve them with the boundary or other conditions imposed by the problem we have in mind. One of the difficulties of plasma physics is that there is no one set of equations which we can write down and identify as the starting point for the theory. Different levels of description and different sets of basic equations are used depending on the problem to be solved.

The difficulties in describing a plasma may perhaps be best explained if we begin with a brief informal outline of the kinetic theory of a neutral gas. In general the behaviour of a gas is very adequately described by a set of equations giving the evolution of its density, velocity, pressure etc., rather than by any kinetic description which contains information about the motion of individual particles. If, however, a kinetic theory is required then it is usually formulated in terms of a particle velocity distribution function $f(\mathbf{r}, \mathbf{v}, t)$ giving the density of particles in a six-dimensional space with coordinates (r, v) at time t. The usual fluid quantities are velocity moments of this, for instance the density is

$$\rho(\mathbf{r}, t) = m \int f(\mathbf{r}, \mathbf{v}, t)\, \mathrm{d}^3 v,$$

with m the mass of a particle, and the fluid velocity is the average particle velocity, that is

$$\boldsymbol{u}(\boldsymbol{r}, t) = \frac{1}{n}\int \boldsymbol{v} f(\boldsymbol{r}, \boldsymbol{v}, t)\, \mathrm{d}^3 v,$$

Here n is the number of particles per unit volume, equal to ρ/m. The evolution of this velocity distribution is described by some kinetic equation, which must be derived from a detailed consideration of the particle dynamics of the system, and which for a neutral gas is Boltzmann's equation. A fundamental property of this equation is that any initial spatially uniform distribution function relaxes towards a Maxwellian

$$f(\boldsymbol{v}) = \frac{n}{(2\pi\kappa T/m)^{3/2}} \exp(-m(\boldsymbol{v}-\boldsymbol{u})^2/2\kappa T), \tag{1.2}$$

where T is the temperature, κ Boltzmann's constant, $\boldsymbol{u}$ the average velocity of the gas and n the density of the particles. The Maxwellian is the velocity distribution function of a system in thermal equilibrium.

The crucial feature of gas dynamics, which allows a fluid rather than a kinetic description to be adequate for most purposes, is that the length scales and time scales over which the system changes are usually much greater than the mean free path and the inter-particle collision time. This means that in the kinetic description the effects of gradients and time variations are small compared with the effects of collisions which push the distribution towards a Maxwellian. In consequence, a systematic perturbation theory can be developed (the Chapman–Enskog method) in which the distribution is, to lowest order, Maxwellian as in (1.2), but with the parameters n, $\boldsymbol{u}$, T allowed to be slowly varying functions of space and time. Equations for n, $\boldsymbol{u}$ and T which are just the usual equations for a fluid can then be derived. The essential feature is that the dominant role of collisions in gas dynamics means that the distribution function is very close to being locally Maxwellian which, in turn, means that the gas is described by the parameters n, $\boldsymbol{u}$ and T.

In a plasma the inter-particle force is the long-range Coulomb force which is weak compared to the strong forces between colliding neutral gas molecules. A particle in a plasma feels the effect of many particles at once and its velocity undergoes a random series of small changes, rather than the sudden changes in velocity produced by neutral gas collisions.

The effect of these Coulomb collisions is again to reduce the velocity distribution to the thermal equilibrium Maxwellian, but an essential feature of plasma physics is that many phenomena of importance take place on time scales much shorter than that associated with the relaxation to thermal equilibrium. For such high-frequency, short-scale phenomena the plasma must be described by a kinetic equation.

Thus, both fluid and kinetic descriptions of a plasma are in regular use, whereas in neutral gas dynamics kinetic theory is only necessary to describe some rather specialized phenomena. Fluid equations are used in plasma physics when large-scale behaviour is being considered, for instance the equilibria of containment systems and their stability against gross motions of the whole plasma. Kinetic equations are used to describe high-frequency wave propagation and the so-called microinstabilities which involve the growth of small-scale perturbations and are driven by the existence of non-Maxwellian velocity distributions. Both types of description will be used in later parts of this book.

1.4 The Debye length

This parameter, which first appeared in the theory of electrolytes, is of fundamental importance in a plasma, and its relation to the average inter-particle spacing plays an important role in determining the strength of Coulomb collisions. If a positive charge, say, is placed in a plasma then electrons are attracted towards it and ions repelled and the effect is to create a screen of negative charge around it. As a result the potential produced by the charge falls off faster than $1/r$ and the Debye length is a characteristic length beyond which the effect of the charge is screened off.

We consider the electrons and suppose that their density is n in the unperturbed system. Then in a potential ϕ the probability of finding an electron with speed v is proportional to $\exp[(-\frac{1}{2}mv^2 + e\phi)/\kappa T]$, this being simply the thermal equilibrium Boltzmann energy distribution. Integrating over velocity we find the electron density is proportional to $\exp(e\phi/\kappa T)$, so that

$$n_e = n_0 \mathrm{e}^{e\phi/\kappa T} \approx n_0(1 + e\phi/\kappa T),$$

and in a potential ϕ the charge density perturbation due to the electrons is $-n_0 e^2\phi/\kappa T$, if $|e\phi|$ is assumed to be small compared to κT. Assuming, for simplicity, that the ions just form a uniform background, then Poisson's equation relates the potential to the charge density and gives

$$\nabla^2\phi = \frac{n_0 e^2 \phi}{\varepsilon_0 \kappa T}.$$

If we look for a spherically symmetrical solution to this, then we have

$$\frac{1}{r^2}\frac{\mathrm{d}}{\mathrm{d}r}\left(r^2\frac{\mathrm{d}\phi}{\mathrm{d}r}\right) = \frac{1}{\lambda^2}\phi,$$

the solution of which tending to zero at infinity is

$$\phi = A\frac{e^{-r/\lambda}}{r},$$

with $\lambda^2 = \varepsilon_0\kappa T/(n_0 e^2)$ and A an arbitrary constant. This gives the behaviour of the potential surrounding a charge, which goes as $\mathrm{e}^{-r/\lambda}/r$ instead of $1/r$

as it would in a vacuum. The effect of the charge is only felt within a distance of order λ, the Debye length.

In a plasma any given particle attracts towards it a screening charge in the same way, with the result that the effective range of the inter-particle force is of the order of the Debye length. A test particle moving through the plasma interacts at any instant with the particles in the Debye sphere, i.e. sphere of radius λ, surrounding it. Any change in its velocity due to 'collisions' with these particles, is the result of a non-zero resultant force on the test particle from the particles within its Debye sphere. The more particles there are within the Debye sphere the more uniformly will they be distributed around the test particle and the less the likelihood of any imbalance producing a force on the test particle. We can thus arrive in an intuitive way at one of the basic results of more elaborate plasma kinetic theories, namely that the parameter which determines the strength of collisions in a plasma and hence the rate of relaxation to thermal equilibrium is the number of particles in the Debye sphere, $n_0\lambda^3$. For magnetically confined fusion plasmas and most space plasmas this is generally a large number, so that collisional effects are weak. The dynamics of the plasma is then not dominated by inter-particle collisions and it can support a variety of waves and instabilities which must be described by kinetic theory. Sometimes this property is taken as part of the definition of a plasma. Systems with few particles in the Debye sphere are called non-ideal or non-Debye plasmas and are a field of specialized study in their own right. They can occur in the compressed plasmas of interest in inertial confinement systems.

The rich variety of phenomena exhibited in a plasma, from large-scale fluid motions to microscopic high frequency oscillations, makes it a challenging and interesting object of study, both for the experimentalist and theorist. Despite the considerable progress which has been made towards its understanding there are still many problems awaiting solution.

2 Motion of a charged particle

2.1 Introduction

It has been pointed out in the introductory chapter that in a hot plasma inter-particle collisions are relatively weak, so that over time scales of interest a particle may remain close to its orbit in the macroscopic fields in the plasma, without being significantly deflected by the microscopic fields arising from other particles. For this reason it is useful to have a knowledge of the behaviour of single particles in electric and magnetic fields. This is not only an essential step towards a more exact analysis of plasma behaviour, but also by itself allows us to understand the basic ideas underlying some magnetic confinement systems.

The usual procedure in developing this orbit theory, and the one which we shall follow here, is to begin with motion in a steady uniform magnetic field, where the exact particle orbit is easily calculated, then use this as the basis for a perturbation theory which will describe the particle orbit in fields varying over suitable long length and time scales. This basic theory is developed in the first part of the chapter and some applications of it discussed in the latter part.

2.2 Motion in a uniform magnetic field

The equation of motion of a particle of mass m and charge q in given electric and magnetic fields is

$$\frac{\mathrm{d}\boldsymbol{v}}{\mathrm{d}t} = \frac{q}{m}(\boldsymbol{E} + \boldsymbol{v} \times \boldsymbol{B}), \tag{2.1}$$

where in the most general case $\boldsymbol{E}$ and $\boldsymbol{B}$ could be functions of both position and time. Unfortunately this general case is not amenable to analytic solution, so to obtain a description of the motion, other than by numerical integration of the equation, we have to resort to various approximation techniques. We begin by discussing the case where $\boldsymbol{E} = 0$, while $\boldsymbol{B}$ is constant, then in subsequent sections extend the description to more general cases.

Taking B along the z-axis, the equations of motion are

$$\frac{dv_x}{dt} = \Omega v_y$$

$$\frac{dv_y}{dt} = -\Omega v_x$$

$$\frac{dv_z}{dt} = 0, \tag{2.2}$$

where $\Omega = qB/m$. The general solution is

$$\begin{aligned} v_x &= v_\perp \cos(\Omega t + \theta) \\ v_y &= -v_\perp \sin(\Omega t + \theta) \\ v_z &= v_\parallel, \end{aligned} \tag{2.3}$$

with $v_\perp$, $v_\parallel$, and θ constants. In the x–y plane the solution simply represents uniform motion in a circle with speed $v_\perp$, the perpendicular (to the magnetic field) velocity component. The constant θ just fixes the position of the particle on the circle at $t = 0$. Superimposed on this circular motion is a uniform parallel velocity, $v_\parallel$, so the final result is motion along a helix. The nature of this motion is easily understood if it is noted that the magnetic force is always perpendicular to the particle velocity, so that it does no work on the particle and the magnitude of the velocity is constant. The force is then of constant magnitude and at right angles to the velocity and the field, just what is needed to produce circular motion in the plane perpendicular to the field. The quantity Ω, the angular frequency of the circular motion, is called the cyclotron frequency, while the radius of the orbit in the x–y plane, given by

$$r_L = \frac{v_\perp}{|\Omega|}$$

is called the Larmor radius.

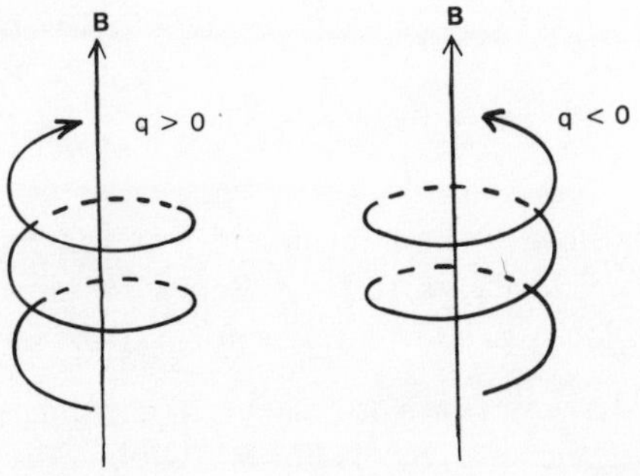

Figure 2.1 Particle orbits in a steady uniform field.

The direction of gyration around the field direction can be deduced from (2.3), or more easily by noting that the magnetic force must be towards the centre of the orbit, the conclusion being that a positively charged particle rotates in a left-handed sense when viewed along the field and a negatively charged particle in the opposite sense. This is illustrated in Figure 2.1.

2.3 Guiding centre drifts

The motion described above can be thought of as gyration about a centre (the guiding centre), which is moving along the direction of the field. In many important applications field gradients or rates of change are such that the particle sees only a small variation during each gyration. Under these circumstances the motion is close to that in a uniform field and can still be thought of as circular motion on which is superimposed a guiding centre drift which may no longer be uniform motion along the field.

The simplest extension is to consider a steady uniform electric field superimposed on the magnetic field of section (2.1). A component along z simply produces a constant acceleration along the z-direction and is of no great interest. Less trivial is a component perpendicular to the magnetic field, let us say in the y-direction, so that our equations of motion are (neglecting the z-component)

$$\frac{\mathrm{d}v_x}{\mathrm{d}t} = \Omega v_y$$

$$\frac{\mathrm{d}v_y}{\mathrm{d}t} = -\Omega v_x + \frac{q}{m}E.$$

These are most easily solved by noting that, putting

$$v_x = \frac{qE}{m\Omega} + V_x$$

transforms them to

$$\frac{\mathrm{d}V_x}{\mathrm{d}t} = \Omega v_y$$

$$\frac{\mathrm{d}v_y}{\mathrm{d}t} = -\Omega V_x,$$

i.e. a set identical to those for a uniform field. Thus we have the motion described before, with a velocity $qE/m\Omega = E/B$ in the x-direction superimposed on it. Writing this velocity in a form independent of the particular coordinate system used here we see that it is

$$\boldsymbol{V}_{\mathrm{E}} = \frac{\boldsymbol{E} \times \boldsymbol{B}}{B^2}. \tag{2.4}$$

This guiding centre drift produced by an electric field perpendicular to a magnetic field is known as the $\boldsymbol{E} \times \boldsymbol{B}$ drift, for reasons evident from (2.4). This solution is exact, so that $\boldsymbol{E} \times \boldsymbol{B}$ drift can, in principle, be arbitrarily large, unlike most of the other drifts which rely on the assumption of small departures from the basic circular orbit.

A simple modification of this calculation shows that if a steady external force $\boldsymbol{F}$ acts on a particle, the most obvious example being a gravitational force, there results an external force drift

$$\boldsymbol{V}_{\mathrm{F}} = \frac{1}{q}\frac{\boldsymbol{F} \times \boldsymbol{B}}{B^2}. \tag{2.5}$$

While we are dealing with the subject of electric fields, we shall consider the effect of a time-dependent electric field, again taken to be along the y-axis so that we have

$$\frac{\mathrm{d}v_x}{\mathrm{d}t} = \Omega v_y$$

$$\frac{\mathrm{d}v_y}{\mathrm{d}t} = -\Omega v_x + \frac{q}{m}E(t).$$

As a first step we take out the $\boldsymbol{E} \times \boldsymbol{B}$ drift by letting

$$v_x = \frac{qE}{m\Omega} + V_x,$$

as before, to obtain

$$\frac{\mathrm{d}V_x}{\mathrm{d}t} = \Omega v_y - \frac{q}{m\Omega}\frac{\mathrm{d}E}{\mathrm{d}t}$$

$$\frac{\mathrm{d}v_y}{\mathrm{d}t} = -\Omega V_x.$$

Because of the time dependence of E this is no longer exactly in the form of the equations for a uniform field, but we note that it now differs from them only by a term $q/m\Omega \cdot \mathrm{d}E/\mathrm{d}t$ which, on the assumption that the time scale for changes in E is much greater than $1/\Omega$, is much less than $(q/m)E$. A similar velocity transformation, namely $v_y = V_y + (q/m\Omega^2)(\mathrm{d}E/\mathrm{d}t)$, eliminates this term from the first of the above equations, giving

$$\frac{\mathrm{d}V_x}{\mathrm{d}t} = \Omega V_y$$

$$\frac{\mathrm{d}V_y}{\mathrm{d}t} = -\Omega V_x - \frac{q}{m\Omega^2}\frac{\mathrm{d}^2E}{\mathrm{d}t^2}.$$

If E is a smooth slowly-varying function then the last term in the equation for $\mathrm{d}V_y/\mathrm{d}t$ will be still smaller than $(q/m\Omega)(\mathrm{d}E/\mathrm{d}t)$ and, if a first approximation to the motion is sought, may be neglected. Thus, to this order of approximation we have a drift $(q/m\Omega^2)(\mathrm{d}E/\mathrm{d}t)$ in the y-direction superimposed on the $\boldsymbol{E} \times \boldsymbol{B}$ drift. In general this drift, the polarization drift, is given by

$$\boldsymbol{v}_{\mathrm{p}} = \frac{q}{m\Omega^2}\frac{\mathrm{d}\boldsymbol{E}}{\mathrm{d}t}. \tag{2.6}$$

Unlike the $\boldsymbol{E} \times \boldsymbol{B}$ drift this drift is in opposite directions for ions and electrons and so gives rise to a current in a plasma, rather than the drift of the entire plasma which results from the $\boldsymbol{E} \times \boldsymbol{B}$ drift. The origin of the name can be seen by considering the effect on a slab of plasma of switching on an electric field in the direction shown in Figure 2.2. As the field is increased, electrons and ions drift in the directions shown and build up a surface charge which remains when the field has reached its final value. This is just like the polarization charge induced in a dielectric and shows that the plasma behaves in some ways like a dielectric.

We shall now turn our attention to non-uniform magnetic fields, always assuming that the variation over distances of the order of the Larmor radius is small. Since the magnetic field is a vector quantity its variation in space is described by the tensor with elements

$$\begin{pmatrix} \dfrac{\partial B_x}{\partial x} & \dfrac{\partial B_x}{\partial y} & \dfrac{\partial B_x}{\partial z} \\[2ex] \dfrac{\partial B_y}{\partial x} & \dfrac{\partial B_y}{\partial y} & \dfrac{\partial B_y}{\partial z} \\[2ex] \dfrac{\partial B_z}{\partial x} & \dfrac{\partial B_z}{\partial y} & \dfrac{\partial B_z}{\partial z} \end{pmatrix} \tag{2.7}$$

Only eight of these nine quantities are independent since the condition $\boldsymbol{\nabla}\cdot\boldsymbol{B} = 0$ implies that the sum of elements on the main diagonal is zero. We shall consider effects due to different magnetic field configurations separately,

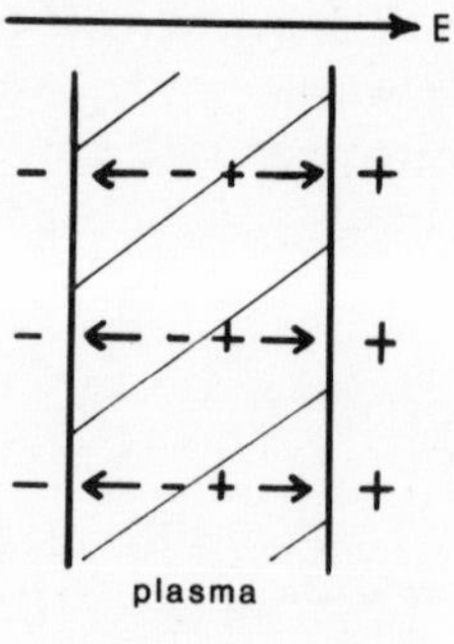

Figure 2.2 Effect of the polarization drift.

merely bearing the above in mind as a check that all possible types of gradient have been looked at. Throughout we shall take the coordinate axes to be such that at the instantaneous position of the particle the magnetic field is along the z-axis.

If we suppose initially that the magnetic field is in the same direction everywhere, but varies in magnitude parallel to this direction, then only B_z is non-zero, and the derivatives $\partial B_z/\partial x$, $\partial B_z/\partial y$ may be non-zero. By choosing the axes suitably we may ensure that only one of these, $\partial B_z/\partial x$ say, is non-zero at the particle's guiding centre. The x and y components of the equation of motion are then

$$\frac{\mathrm{d}v_x}{\mathrm{d}t} = \Omega(x)v_y \approx (\Omega + \Omega' x)v_y \tag{2.8}$$

$$\frac{\mathrm{d}v_y}{\mathrm{d}t} = -\Omega(x)v_x \approx -(\Omega + \Omega' x)v_x,$$

where the cyclotron frequency, which is a function of x, has been expanded in a Taylor series, with only the first two terms kept. The quantities Ω and Ω' are $\Omega(x)$ and its derivative evaluated at the centre of the orbit, while x is measured from this position. The terms involving Ω' in the above equations are small in comparison with the other terms, so in them we substitute for x and the velocities the values corresponding to the basic circular motion. This neglects terms involving the product of Ω' and corrections to this motion on the basis that these will be smaller still. In this way we obtain the following approximation to (2.8):

$$\frac{\mathrm{d}v_x}{\mathrm{d}t} = \Omega v_y - \frac{v_\perp^2}{\Omega}\Omega' \sin^2(\Omega t + \theta) \tag{2.9}$$

$$\frac{\mathrm{d}v_y}{\mathrm{d}t} = -\Omega v_x - \frac{v_\perp^2}{\Omega}\Omega' \sin(\Omega t + \theta)\cos(\Omega t + \theta).$$

If we now note that a particle in a magnetic field does not gain energy, then we may require that the acceleration, averaged over a cyclotron period, must be zero. Taking such an average over (2.9) we obtain

$$0 = \Omega\langle v_y\rangle - \tfrac{1}{2}\Omega' \frac{v_\perp^2}{\Omega}$$

$$0 = -\Omega\langle v_x\rangle,$$

so we conclude that there must be an average drift velocity $1/2(v_\perp^2\Omega'/\Omega^2)$ in the y-direction. In a form independent of the coordinate system this gradient drift is

$$\boldsymbol{V}_\mathrm{G} = \frac{1}{2}\frac{qv_\perp^2}{m}\frac{\boldsymbol{B}\times\nabla B}{\Omega^2 B}. \tag{2.10}$$

Readers of a mathematical turn of mind may care to note that the above procedure is essentially a perturbation expansion in Ω' with the drift determined by the condition that secular terms do not appear. The physical reason for the drift is illustrated in Figure 2.3, showing the orbit of a particle which rotates in a clockwise direction in absence of a gradient. The result of the gradient is that the radius of curvature of the orbit is slightly less when it is in the higher field regions, so the orbit is not quite closed and we have the effect shown in exaggerated form in the figure. Particles of opposite sign drift in opposite directions.

Now let us look at a system in which the magnetic field lines are curved. If we take the curvature in the x–z plane as shown in Figure 2.4, then we can see that $\partial B_x/\partial z$ is non-zero at the origin, and that the radius of curvature is given by noting that

$$\delta\theta \approx \frac{\delta z}{R} \approx \left(\frac{\partial B_x}{\partial z}\,\delta z\right)\bigg/ B$$

so that

$$\frac{1}{R} = \frac{\partial B_x/\partial z}{B}.$$

In general, if $\boldsymbol{R}$ is the vector radius of curvature shown in Figure 2.4, then $\boldsymbol{R}/R^2$ is the perpendicular component of

$$\frac{1}{B^2}(\boldsymbol{B}\cdot\boldsymbol{\nabla})\boldsymbol{B}. \tag{2.11}$$

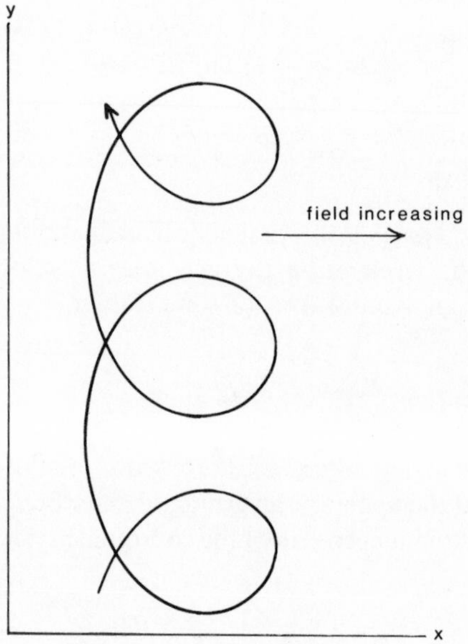

Figure 2.3 Drift in a field gradient.

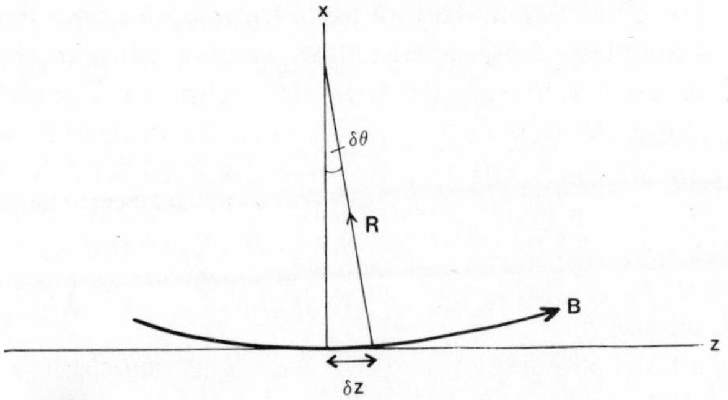

Figure 2.4 Geometry of curved field line.

Having sorted out the geometry we now look at the dynamics of the particle. The simplest way to see the effect of curvature is to note that if the particle is following its guiding centre orbit along the field line it experiences a centrifugal force given by

$$\boldsymbol{F} = -\frac{mv_{\parallel}^2 \boldsymbol{R}}{R^2}$$

and that this force has the same effect as an external force, namely it produces a drift $(1/q)[(\boldsymbol{F} \times \boldsymbol{B})/B^2]$. The curvature drift is thus found to be

$$v_c = \frac{mv_{\parallel}^2}{q} \frac{\boldsymbol{B} \times (\boldsymbol{B} \cdot \boldsymbol{\nabla})\boldsymbol{B}}{B^4}.$$

If the terms $\partial B_y/\partial x$ or $\partial B_x/\partial y$ of the magnetic field gradient tensor are non-zero the effect on the geometry of the field is to produce a twist or shear about the central direction. It can be shown (see Problem 2.2) that such effects do not produce any steady drift motion but simply a distortion of the shape of the particle orbit.

The possible effects which we have not looked at are those due to a time-dependent magnetic field and to non-zero terms on the main diagonal of the gradient tensor. We shall devote a separate section to these since they introduce a new concept, that of an adiabatic invariant.

2.4 Adiabatic invariants: the magnetic moment

Beginning with the first of the two effects mentioned at the end of the previous chapter, we consider a spatially uniform magnetic field $\boldsymbol{B} = B(t)\boldsymbol{z}$. The effect of such a changing magnetic field is to produce an electric field such that, by Faraday's law,

$$\oint \boldsymbol{E} \cdot \mathrm{d}\boldsymbol{l} = -\frac{\mathrm{d}\Phi}{\mathrm{d}t}, \tag{2.12}$$

where the integral is around any closed loop and Φ is the magnetic flux through this loop. The effect on a particle gyrating in the magnetic field is that it is accelerated by the electric field. The orbit of a positive particle is such that increasing Φ causes it to gain energy since its direction of rotation is in what is conventionally the negative direction in the line integral in (2.12). The energy gained in a cyclotron period is

$$\Delta W_{\perp} = -\oint q\boldsymbol{E}\cdot \mathrm{d}\boldsymbol{l} = q\frac{\mathrm{d}\Phi}{\mathrm{d}t}$$

$$= q\pi r_{\mathrm{L}}^2 \frac{\mathrm{d}B}{\mathrm{d}t},$$

where r_{L} is the Larmor radius and the subscript $\perp$ indicates that the change in energy results from a change in the perpendicular velocity, while the parallel velocity and its associated energy are constant. The cyclotron period is $2\pi/\Omega$, so

$$\Delta W_{\perp} \approx \frac{\mathrm{d}W_{\perp}}{\mathrm{d}t}\frac{2\pi}{\Omega} = q\pi \frac{v_{\perp}^2}{\Omega^2}\frac{\mathrm{d}B}{\mathrm{d}t},$$

from which we obtain, on using $W_{\perp} = \frac{1}{2}mv_{\perp}^2$,

$$\frac{1}{W_{\perp}}\frac{\mathrm{d}W_{\perp}}{\mathrm{d}t} = \frac{1}{B}\frac{\mathrm{d}B}{\mathrm{d}t},$$

which integrates to give $W_{\perp}/B = \text{constant}$. The quantity $W_{\perp}/B$ is not an exact constant of the motion, but is a good approximation to a constant in the limit when changes are small in one cyclotron period. Such a quantity is known as an adiabatic invariant. This rather vague definition will suffice for us, but can be improved on by a more elaborate treatment which includes estimates of just how constant the adiabatic invariant is.

Before discussing this adiabatic invariant further, we shall show how it arises in a somewhat different context, namely in an inhomogeneous field where the field strength varies along the field direction. This implies with our usual geometry that $\partial B_z/\partial z \neq 0$, so that, since $\nabla \cdot \boldsymbol{B} = 0$ at least one other term in the main diagonal of the gradient tensor must be non-zero.

The field configuration is illustrated in Fig. 2.5, drawn for $\partial B_z/\partial z < 0$.

The physical effect of such a field is that at the position of the particle there is a radial field component which produces a force along the z-direction given by

$$F_z = qv_{\perp}B_r, \tag{2.13}$$

B_r being the field component directed radially outwards. Now, let us average over a cyclotron period. To do this note that $\boldsymbol{B}_r = \boldsymbol{B}\cdot\hat{\boldsymbol{r}}$, with $\hat{\boldsymbol{r}}$ a unit vector directed from the centre of the orbit towards the particle.

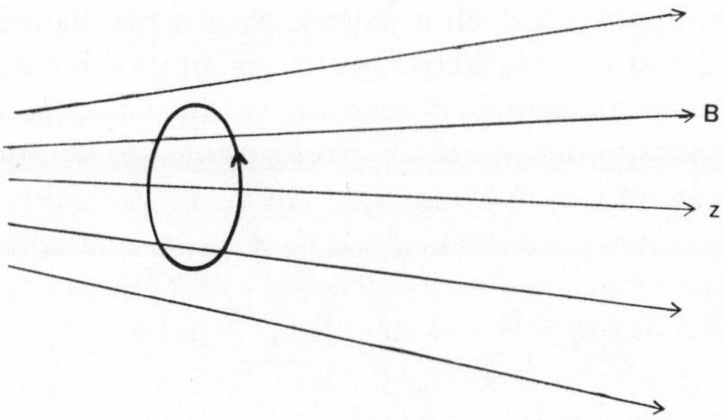

Figure 2.5 Particle orbit and field lines with $\partial B_z/\partial z < 0$.

Thus

$$B_r = B_x \frac{x}{r_L} + B_y \frac{y}{r_L}$$

and if we make the approximations $B_x \approx (\partial B_x/\partial x)x$, $B_y \approx (\partial B_y/\partial y)y$, with the partial derivatives evaluated at the guiding centre, we obtain

$$B_r \approx \frac{\partial B_x}{\partial x}\frac{x^2}{r_L} + \frac{\partial B_y}{\partial y}\frac{y^2}{r_L},$$

which on taking the average around the orbit gives

$$\bar{B}_r \approx \tfrac{1}{2} r_L \left(\frac{\partial B_x}{\partial x} + \frac{\partial B_y}{\partial y} \right) = -\tfrac{1}{2} r_L \frac{\partial B_z}{\partial z} \approx -\tfrac{1}{2} r_L \frac{\partial B}{\partial z}$$

where, in the last step, we use the fact that the z-component of the field is essentially the total field magnitude.

Then, from (2.13), we obtain

$$F_z = -\frac{1}{2} q \frac{v_\perp^2}{\Omega}\frac{\partial B}{\partial z} = -\frac{W_\perp}{B}\frac{\partial B}{\partial z}$$

for the force averaged over a cyclotron period. Now $F_z = (\mathrm{d}/\mathrm{d}z)(\tfrac{1}{2}mv_z^2)$, and so since no work is done on the particle by a steady magnetic field, with the consequence that $\tfrac{1}{2}mv_z^2 + \tfrac{1}{2}mv_\perp^2$ is constant, we have

$$\frac{\mathrm{d}}{\mathrm{d}z} W_\perp = \frac{W_\perp}{B}\frac{\partial B}{\partial z},$$

which integrates to $W_\perp/B = \text{constant}$. Thus we again have adiabatic invariance of the quantity $W_\perp/B$, the only difference being that now B changes because the guiding centre of the particle moves into regions of different field strength as it follows a field line, rather than because the field is time-dependent.

The quantity $W_\perp/B$ can be given a physical meaning if we remember that the magnetic moment associated with a current loop is the current times the area of the loop. If the current carried by the gyrating particle is thought of as being smoothed out around the orbit its value is $qv_\perp/2\pi r_L$, while the area of the orbit is πr_L^2. Taking the product of these gives just $W_\perp/B$, so our result can be expressed by saying that the magnetic moment, generally denoted by μ, associated with a particle in a magnetic field is an adiabatic invariant. If a particle moves towards a region of higher field strength, then energy goes into the perpendicular degrees of freedom at the expense of the parallel velocity. If the field becomes sufficiently strong the parallel velocity goes to zero and the particle is reflected.

Adiabatic invariants can be found quite generally for oscillatory motion in which some parameter of the system is slowly changing, the earliest such system to be studied being a pendulum whose string is being slowly lengthened or shortened. There are other adiabatic invariants which are important in certain plasma physics applications, but we shall not investigate these further. The magnetic moment is the most important and commonly used adiabatic invariant.

2.5 The magnetization current

To the order of accuracy used to analyse the various effects described above they are all additive, so that we can use our results to discuss the behaviour of individual particles and of plasmas in complicated field geometries. Drifts in which the ions and electrons go in opposite directions give rise to a current in an overall neutral plasma where both species are present. However, in describing the behaviour of such a system it is important to realize that even without drifts a current may flow and that this current must be added to that produced by particle drifts. The current in question is precisely analogous to the effective surface and volume currents produced in any magnetized material, and in plasma their physical origin may be seen by referring to

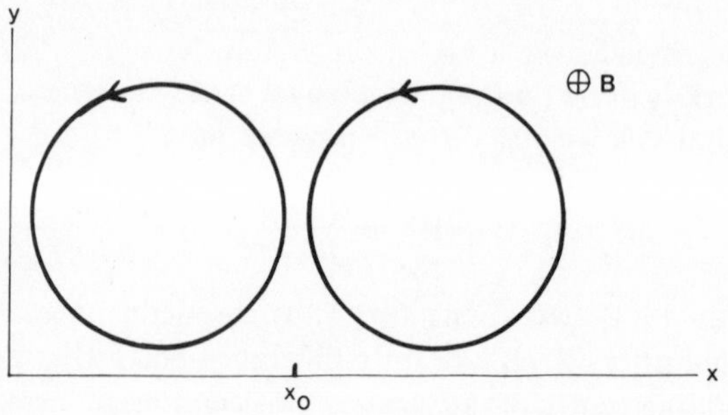

Figure 2.6 Ion orbits in a plane perpendicular to the magnetic field. Electrons rotate in the opposite direction.

Figure 2.6, which is a view of particle orbits in a plane perpendicular to the magnetic field.

If we consider the current at the position x_0 it can be seen that both ions and electrons with orbits centred to the left of x give a current in the positive y-direction, while those with orbits centred to the right give currents in the opposite direction. In a uniform plasma these currents cancel, but if there is a density or temperature gradient there will be an imbalance in the number or average velocity of particles in these two groups with the result that a net current, known as the magnetization current, flows in the plasma.

A calculation of the current density resulting from this mechanism can be made by considering the current through a surface bounded by some closed curve as illustrated in Figure 2.7.

If each of the particles is regarded as being equivalent, on a time scale much longer than the cyclotron period, to a current loop of vector area $\boldsymbol{A}$ carrying a current I, then the only contribution to the total current through the surface is that given by those loops which are intersected by the boundary. If N is the density of these loops then the total current through the surface is

$$\oint IN\boldsymbol{A}\cdot d\boldsymbol{l} = \oint \boldsymbol{M}\cdot d\boldsymbol{l},$$

where $\boldsymbol{M}$ is the total magnetic moment per unit volume. Thus if $\boldsymbol{J}_M$ is the magnetization current density we have

$$\int_S \boldsymbol{J}_M\cdot d\boldsymbol{S} = \oint \boldsymbol{M}\cdot d\boldsymbol{l} = \int_S \nabla\times\boldsymbol{M}\cdot d\boldsymbol{S},$$

from which we conclude that

$$\boldsymbol{J}_M = \nabla\times\boldsymbol{M}. \tag{2.14}$$

The vector magnetic moment associated with a particle is $-(W_\perp/B)\hat{\boldsymbol{b}}$ where $\hat{\boldsymbol{b}}$ is a unit vector along the field and the origin of the $-$ sign is to be found by looking at the direction of rotation of a particle. In a real plasma particles will not all have the same energy, so that in finding the magnetic moment per unit volume an average has to be taken. The resulting current has to be added to that originating in particle drifts.

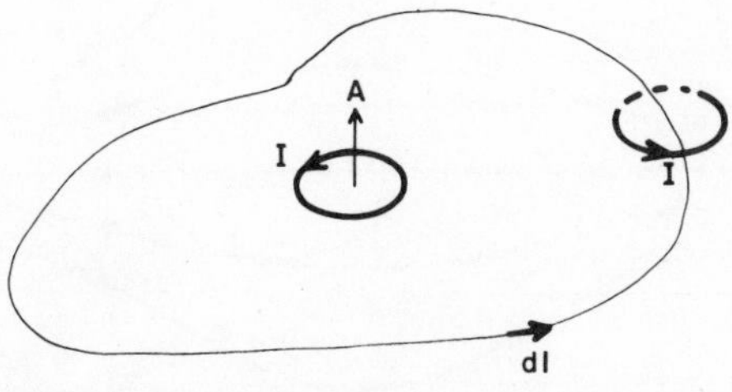

Figure 2.7 Calculation of magnetization current.

2.6 Some applications

Some of the most important consequences of the preceding theory arise from the adiabatic invariance of the magnetic moment and the fact that particles are reflected from regions of high field. This leads to the idea of the type of confinement device known as a mirror machine, the essential feature of which is a magnetic field configuration like that of Figure 2.8, so that charged particles introduced into the central region may be reflected from the regions of stronger magnetic field at each end. If B_0 and B_1 are the fields at the centre and at the ends respectively, then a particle with perpendicular energy $W_\perp^{(0)}$ at the centre will have perpendicular energy $W_\perp^{(0)} B_1/B_0$ if it reaches the end. Since total energy is conserved, the condition that the particle may pass through the high field region and escape is

$$W_\perp^{(0)} \frac{B_1}{B_0} < W = W_\perp^{(0)} + W_\parallel^{(0)},$$

where W is the total energy of the particle and $W_\parallel^{(0)}$ the energy associated with its parallel motion at the centre of the device. Thus a particle escapes if

$$\frac{W_\parallel^{(0)}}{W_\perp^{(0)}} > \frac{B_1 - B_0}{B_0}. \tag{2.15}$$

This condition defines a cone in velocity space, known as the loss cone, with its axis aligned along the magnetic field. If a plasma is created in the centre of the machine particles in the loss cone are not confined. In the absence of interactions between the particles the remainder of the plasma would be confined, but collisions scatter particles into the loss cone and eventually all of the plasma escapes. This situation is aggravated by the existence of loss cone instabilities resulting from the non-Maxwellian nature of the particle velocity distribution after the loss cone particles escape. These instabilities produce enhanced electric field fluctuation levels and increase the rate of scatter of confined particles into the loss cone. The result of these effects is that a simple mirror machine is not a useful containment device, though a

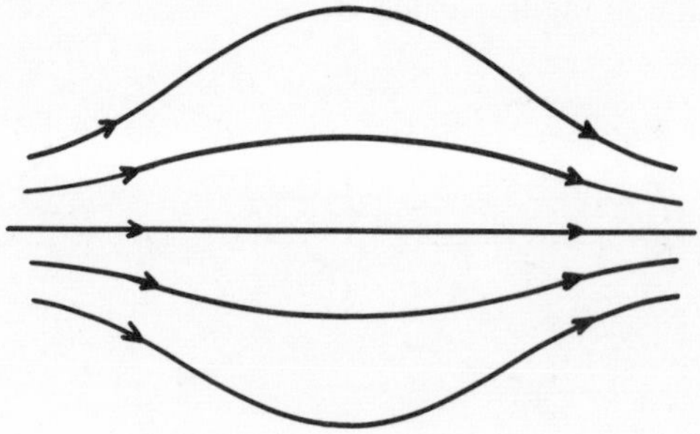

Figure 2.8 Field in mirror machine.

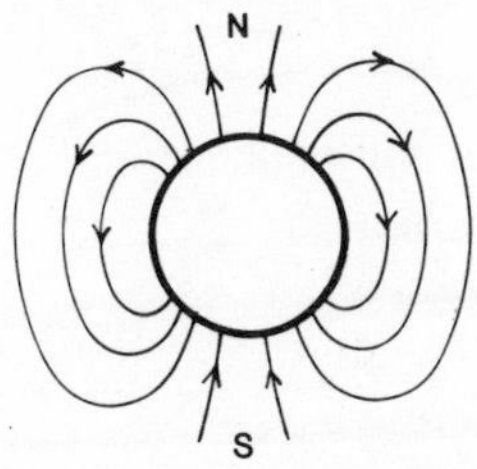

Figure 2.9 The Earth's magnetic field.

more complicated variant known as a tandem mirror is still the subject of some interest. This machine has several field maxima and minima along its axis and relies on electrostatic potential differences to contain the loss cone particles.

Before leaving the subject of mirror machines we should perhaps point out that a plasma confined in a field configuration like that of Figure 2.8 is subject to fluid instabilities, and that in practice a more complicated three-dimensional field structure with the field lines convex towards the plasma is required.

The same principle governs the trapping of particles in the Earth's magnetic field to produce the Van Allen radiation belts of charged particles found to lie above the equatorial regions. The dipole field of the Earth, illustrated in Figure 2.9, is such that particles following a field line may be reflected from the polar regions and trapped for long periods in the low-density plasma of the radiation belts. Particles entering the Earth's field which are not trapped are channelled to earth in the polar regions and are responsible for the aurorae.

Since the ends of a linear containment device cannot be completely plugged by means of magnetic mirrors, an obvious step is to eliminate this problem by a toroidal geometry. Most magnetic containment devices of current interest are of this type, so we shall consider particle orbits in such systems in some detail.

First, let us suppose that we have a simple toroidal field, as shown in Figure 2.10. From Ampère's law it follows that the field strength must

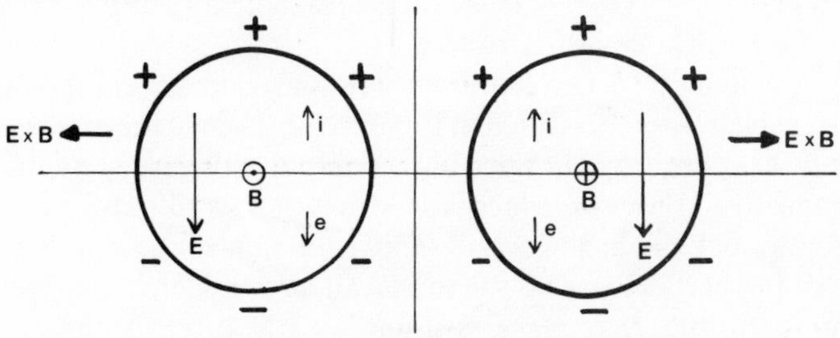

Figure 2.10 Drifts in a toroidal field.

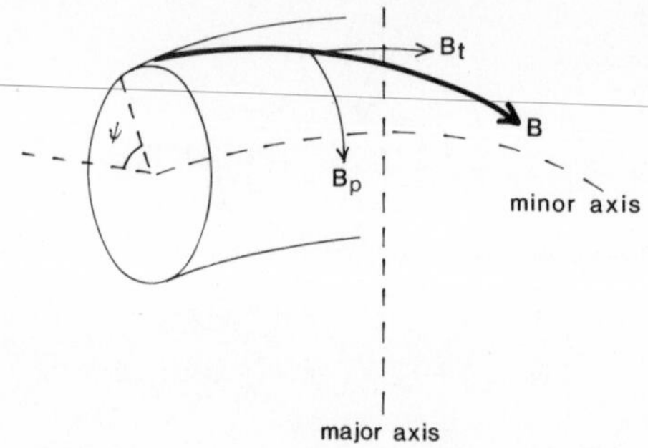

Figure 2.11 Field with rotational transform.

decrease from the inside to the outside of the torus, so that a particle has both gradient and curvature drifts. These are in the same direction and, in the geometry of Figure 2.10, cause ions to drift upwards and electrons downwards. The charge separation which results produces an electric field and the resulting $\boldsymbol{E} \times \boldsymbol{B}$ drift of the whole plasma is such as to produce an outward expansion as shown. We must conclude that a simple toroidal field cannot confine a plasma and look for a more complicated field which will.

If lines of force are twisted around the minor axis of the torus as shown in Figure 2.11 then particles can move freely along lines of force from the upper to the lower part of the torus, so short-circuiting the electric field produced in a purely toroidal system.

The amount of twist in the field is usually measured by the rotational transform, which is the change in the angle ψ shown in Figure 2.11 when the field line goes once around the major axis of the torus. The field may be considered as the superposition of a toroidal field B_t and a poloidal field B_p, in the directions shown. The toroidal component can be produced by external windings forming a simple toroidal solenoid, while the poloidal component can be generated by a current flowing around the torus in the plasma, as in a tokamak, or by extra external windings, as in a stellarator. In a tokamak, which is currently the device in which most effort (and money) has been invested, the poloidal field is small compared to the toroidal field and the rotational transform is less than 2π.

In view of the importance of the tokamak we shall say a little more about the motion of a particle in such a system.

We take cylindrical polar coordinates (R, ϕ, z) and project the motion on to the $(R$–$z)$ plane, as represented in Figure 2.12. Then, if the motion followed a magnetic field line it might trace out a circle as represented by the full line. (This magnetic surface, on which a field line lies, is not always circular in cross-section, but the behaviour is similar for non-circular surfaces.) If the velocity of the particle around the torus can be considered constant then its basic motion in the $(R$–$z)$ plane as it follows a field line is uniform circular motion of angular frequency ω, say. On to it is, however, superimposed the

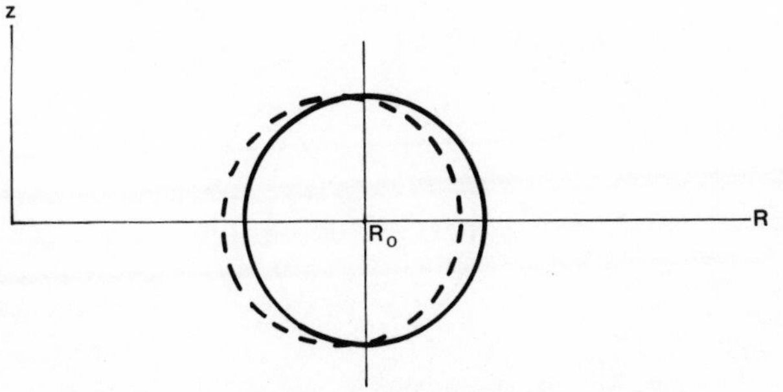

Figure 2.12 Untrapped particle orbit in a tokamak.

drift in the z-direction discussed above and which we assume, for simplicity, to be constant.

With these assumptions the motion of the particle in the $(R\text{–}z)$ plane is described by the equations

$$\frac{\mathrm{d}R}{\mathrm{d}t} = -\omega z$$

$$\frac{\mathrm{d}z}{\mathrm{d}t} = \omega(R - R_0) + v_d,$$

the integral of which is

$$\left(R - R_0 + \frac{v_d}{\omega}\right)^2 + z^2 = \text{constant}. \tag{2.16}$$

This represents a circle whose centre is displaced towards the centre of the torus a distance v_d/ω from the centre of the magnetic surfaces, as indicated by the dotted line in Figure 2.12. For particles which can travel right around the torus in this way the deviation from the magnetic surface is generally quite small.

There is, however, another class of particles which are trapped in one section of the torus. These occur because, as we have pointed out, the magnetic field strength is greater at the inner edge than at the outer edge of the torus, and so a particle in the outer region may be reflected as it moves along the magnetic field towards regions of stronger field.

Referring to Figure 2.13, the particle orbit projected on to the $R\text{–}z$ plane would, in the absence of toroidal drift effects, oscillate along the circular flux surface between the angles $\pm\,\theta_0$ in the coordinate system shown. If there is an upward drift v_d, then

$$\dot{r} = v_d \sin\theta,$$

and if we take a particle which starts at its reflection point where $\theta = -\,\theta_0$, its distance from the centre of the flux surface decreases until it cuts the axis

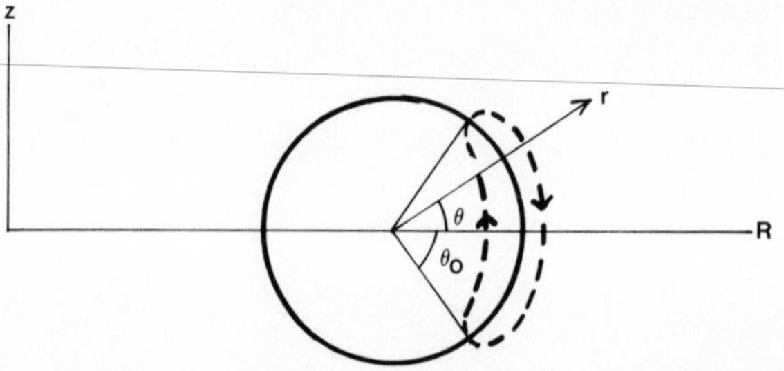

Figure 2.13 Banana orbit of trapped particle in a tokamak.

$\theta = 0$, then increases until it reaches its other reflection point at $+\theta_0$. Then, as the particle comes back, r first increases then decreases. The result is that the projection of the orbit on to the (R–z) plane is as shown in Figure 2.13. This type of orbit is known, for obvious reasons, as a banana orbit. The existence of trapped particles with this behaviour is of great importance in theories attempting to explain the transport of energy and particles due to collisional effects in a torus.

Finally in this section we shall illustrate the use of single-particle orbit theory to analyse the behaviour of a plasma, deriving conditions for an equilibrium in which the plasma currents are consistent with the magnetic field, these quantities being, of course, related through Maxwell's equations. We consider a system with cylindrical symmetry in which, with respect to cylindrical polar coordinates (R, ϕ, z), the magnetic field is in the z-direction and depends only on R. A particle then drifts around in the azimuthal direction as a result of the gradient drift and we have (from 2.10)

$$v_\phi = \frac{1}{2}\frac{q}{m}\frac{v_\perp^2}{\Omega^2}\frac{\mathrm{d}B}{\mathrm{d}R} = \frac{1}{2}\frac{m}{q}\frac{v_\perp^2}{B^2}\frac{\mathrm{d}B}{\mathrm{d}R}.$$

We shall suppose that there are n ions of charge e and n electrons of charge $-e$ per unit volume. If we also assume that both species have temperature T, then the average value of $\frac{1}{2}mv_\perp^2$ for each species is κT, where κ is Boltzmann's constant, and $n\kappa T$ is the partial pressure of each species. With these assumptions the current resulting from the gradient drift is

$$J_{\phi G} = \frac{P}{B^2}\frac{\mathrm{d}B}{\mathrm{d}R},$$

where P is the total plasma pressure. The only other contribution to the current is the magnetization current,

$$J_{\phi M} = \frac{\mathrm{d}}{\mathrm{d}R}\left(\frac{P}{B}\right)$$

so that the total current is

$$J_\phi = J_{\phi G} + J_{\phi M} = \frac{1}{B}\frac{\mathrm{d}P}{\mathrm{d}R}. \tag{2.17}$$

In the present geometry, Ampère's Law $\boldsymbol{\nabla} \times \boldsymbol{B} = \mu_0 J$ becomes

$$\frac{\mathrm{d}B}{\mathrm{d}R} = -\mu_0 J_\phi. \tag{2.18}$$

If we now look for a self-consistent equilibrium in which (2.17) and (2.18) are both true, then we must have

$$\frac{\mathrm{d}}{\mathrm{d}R}\left(P + \frac{B^2}{2\mu_0}\right) = 0. \tag{2.19}$$

We shall see in the next chapter that in a treatment of the plasma as a conducting fluid the force exerted on the plasma by the magnetic field is as though the field had a pressure $B^2/2\mu_0$. Thus (2.19) can be interpreted as saying that in equilibrium the pressure must be uniform throughout the system, the total pressure being the sum of magnetic and particle pressures.

It is interesting to note that a model which considers individual non-interacting particles gives the same results as a fluid model, since, as pointed out in Chapter 1, we might have expected the validity of the latter to depend on there being strong collisional effects. However, when the approximations of this chapter are valid, particles cannot travel freely across field lines but are (except on the slow time scale associated with drifts) constrained within the dimensions of a Larmor radius, which is much smaller than the gradient scale lengths. So far as variation across field lines is concerned, this tying of particles to field lines plays a role similar to that of collisions, and fluid equations are a good approximation.

Problems

2.1 A slab of plasma is placed in a magnetic field and an electric field applied in the direction perpendicular to the magnetic field as shown. If the electric field is

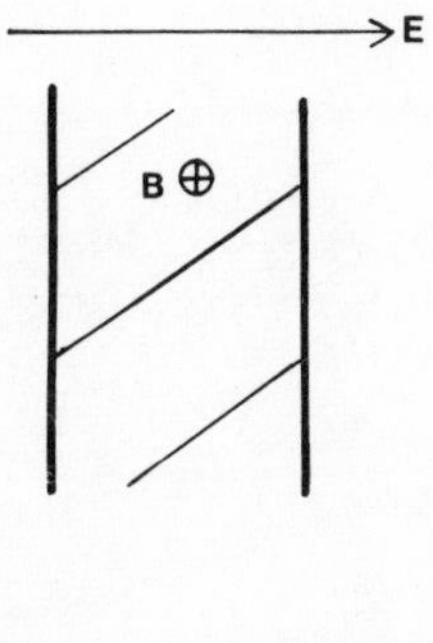

increased slowly from 0 to a final value E_0 outside the plasma, show, by considering build-up of charge on the plasma surface due to the polarization drift, that the final field inside the plasma is

$$E_0/(1+\rho/B^2\varepsilon_0),$$

where ρ is the plasma density.

2.2 Consider the orbit of a particle in a sheared magnetic field with $\boldsymbol{B}$ along the z-direction at the guiding centre and $\partial B_x/\partial y=0$. Show that there is no guiding centre drift.

2.3 A magnetic mirror is formed by two coaxial single-turn coils of radius a, a distance $2a$ apart, each carrying the same current in the same direction. Show that a particle which starts on the axis midway between the coils will be trapped if its initial velocity components satisfy

$$v_\parallel^2<\left[\sqrt{2}\frac{5^{3/2}+1}{5^{3/2}}-1\right]v_\perp^2,$$

the Larmor radius being assumed to be much less than a.

2.4 In a tokamak the magnetic field decreases with major radius in such a way that $\mathrm{d}B/\mathrm{d}R\approx B/R$. If the particle velocity distributions are isotropic, show that the average gradient and curvature drifts are equal.

3 Low-frequency phenomena

3.1 Introduction

The object of this chapter is to examine some of the methods used to describe a plasma in the regime where collisional effects are important. For the most part we shall be looking at fluid theories, which give a good description of large-scale slow phenomena in a plasma, but are inadequate to describe the high-frequency waves and instabilities which are the subject of the two chapters following. After an introduction setting out the basic equations of magnetohydrodynamics we go on to look at fluid equilibria and their stability and at the Alfvén and magnetosonic waves which are predicted by the theory. Finally we give a brief outline of the kinetic equations which describe collisional effects and the way in which fluid equations can be obtained from them and transport coefficients like resistivity or thermal conductivity calculated.

3.2 Magnetohydrodynamics

Magnetohydrodynamics is basically classical fluid dynamics with the additional complication that the fluid is assumed to be electrically conducting and so capable of generating and interacting with magnetic fields. The fluid is assumed to be a sufficiently good conductor that any electric charge leaks away in a time short compared with those of interest. Thus it is electrically neutral and the only electric fields in the system are those generated by varying magnetic fields.

If the fluid has density ρ and velocity $\boldsymbol{v}$, each of these being functions of position and time, then the first of the equations which we require is that expressing conservation of mass, namely

$$\frac{\partial \rho}{\partial t} + \boldsymbol{\nabla}\cdot(\rho\boldsymbol{v}) = 0. \tag{3.1}$$

If this equation is integrated over a finite volume and the second term converted to a surface integral by means of Gauss's theorem then it expresses the simple fact that the rate of change of mass inside the surface equals the mass flow through the boundary.

Next we require a momentum equation, of the form

$$\rho\frac{\partial \boldsymbol{v}}{\partial t}+\rho(\boldsymbol{v}\cdot\boldsymbol{\nabla})\boldsymbol{v}=-\boldsymbol{\nabla} p+\boldsymbol{J}\times\boldsymbol{B}. \tag{3.2}$$

The convective derivative $\partial/\partial t+\boldsymbol{v}\cdot\boldsymbol{\nabla}$ gives the rate of change of any quantity as it moves with the fluid, so the left-hand side of (3.2) is just the total rate of change of momentum of a fluid element, while the right-hand side is the sum of the forces due to the fluid pressure p and the magnetic field. To determine the pressure there are a number of possibilities. If the fluid is assumed to be incompressible (though not necessarily of uniform density throughout), then the rate of change of density following a fluid element is zero, i.e.

$$\left(\frac{\partial}{\partial t}+\boldsymbol{v}\cdot\boldsymbol{\nabla}\right)\rho=0, \tag{3.3}$$

which combined with (3.1) gives

$$\boldsymbol{\nabla}\cdot\boldsymbol{v}=0. \tag{3.4}$$

Alternatively, compression of the fluid may be assumed to be adiabatic, so that for any element of fluid (p/ρ^{γ}) is a constant and

$$\left(\frac{\partial}{\partial t}+\boldsymbol{v}\cdot\boldsymbol{\nabla}\right)(p/\rho^{\gamma})=0, \tag{3.5}$$

γ, the ratio of specific heats, being generally taken to have its monatomic gas value 5/3. An energy equation, taking into account thermal transport processes, may be needed if this is not a sufficiently good approximation.

As well as these fluid equations we need equations to describe the field and current. In magnetohydrodynamics we are generally dealing with low-frequency effects in which the displacement current is negligible, so that

$$\boldsymbol{\nabla}\times\boldsymbol{B}=\mu_0\boldsymbol{J}. \tag{3.6}$$

The current is determined by Ohm's law, which, since the electric field in a frame of reference moving with the fluid is $\boldsymbol{E}+\boldsymbol{v}\times\boldsymbol{B}$, takes the form

$$\boldsymbol{E}+\boldsymbol{v}\times\boldsymbol{B}=\eta\boldsymbol{J}, \tag{3.7}$$

with η the resistivity of the plasma. A more complicated form of Ohm's law is obtained if additional physical effects like the Hall or thermoelectric effects are included. The electric and magnetic fields are related through Faraday's law:

$$\boldsymbol{\nabla}\times\boldsymbol{E}=-\frac{\partial \boldsymbol{B}}{\partial t}. \tag{3.8}$$

Equations (3.1), (3.2), (3.5), (3.6), (3.7) and (3.8) give a complete set of equations describing adiabatic motions in a conducting fluid. Combining (3.6)–(3.8) gives

$$\frac{\partial \boldsymbol{B}}{\partial t} = \frac{\eta}{\mu_0} \nabla^2 \boldsymbol{B} + \nabla \times (\boldsymbol{v} \times \boldsymbol{B}). \tag{3.9}$$

If L is a characteristic length associated with the field gradients and U a characteristic fluid velocity, then the ratio of the magnitude of the second term on the right-hand side of (3.9) to the first is approximately

$$\mu_0 \frac{UL}{\eta},$$

a dimensionless quantity known as the magnetic Reynolds number R. If the conductivity of the plasma is low, so that $R \ll 1$, then the magnetic field is dominated by the effect of resistivity and (3.9) reduces to a diffusion equation, describing the diffusion of magnetic field into a conductor, the motion of which is not significant. However, it is more usual for the plasma to have a high conductivity, so that $R \gg 1$ and the second term in (3.9) is dominant. Often a good description can be obtained by letting the resistivity go to zero, so that (3.7) becomes

$$\boldsymbol{E} + \boldsymbol{v} \times \boldsymbol{B} = 0. \tag{3.10}$$

In this approximation, known as ideal magnetohydrodynamics, there is no resistive dissipation and the magnetic field satisfies

$$\frac{\partial \boldsymbol{B}}{\partial t} = \nabla \times (\boldsymbol{v} \times \boldsymbol{B}). \tag{3.11}$$

If Φ is the magnetic flux through a surface S bounded by a curve C fixed in space, then integrating (3.11) over S and using Stokes's theorem gives

$$\frac{\mathrm{d}\Phi}{\mathrm{d}t} = \oint_C (\boldsymbol{v} \times \boldsymbol{B}) \cdot \mathrm{d}\boldsymbol{l}. \tag{3.12}$$

The right-hand side is just the change in flux to be expected if we imagine magnetic lines of force being convected across the boundary of S with the fluid velocity. This brings us to the important idea that in ideal magnetohydrodynamics the magnetic field can be regarded as being 'frozen into' the fluid, magnetic flux being carried along with fluid velocity.

If (3.6) is used to eliminate J from the right-hand side of (3.2) the result is

$$\rho \frac{\partial \boldsymbol{v}}{\partial t} + \rho (\boldsymbol{v} \cdot \nabla) \boldsymbol{v} = -\nabla \left(P + \frac{B^2}{2\mu_0} \right) + \frac{1}{\mu_0} (\boldsymbol{B} \cdot \nabla) \boldsymbol{B}. \tag{3.13}$$

From (3.13) we see that the magnetic fluid produces what is effectively an extra contribution to the pressure, the magnetic pressure $B^2/2\mu_0$. The other part of the magnetic force $(1/\mu_0)(\boldsymbol{B} \cdot \nabla)\boldsymbol{B}$ is non-zero if there are curved or divergent lines of force. Qualitatively its effect is as though the magnetic lines of force were in tension, and the force due to this tension transferred to the plasma. These ideas of magnetic pressure and tension are useful in trying to visualize the behaviour of plasma in a magnetic field.

3.3 MHD equilibria

If we look for a static equilibrium in which velocities and time derivatives vanish, then we obtain the magnetostatic equations

$$\begin{aligned} \nabla p &= \boldsymbol{J} \times \boldsymbol{B} \\ \nabla \times \boldsymbol{B} &= \mu_0 \boldsymbol{J} \\ \nabla \cdot \boldsymbol{B} &= 0. \end{aligned} \tag{3.14}$$

An immediate consequence of these is that

$$\boldsymbol{B} \cdot \nabla p = \boldsymbol{J} \cdot \nabla p = 0, \tag{3.15}$$

that is, the field and current are both perpendicular to the pressure gradient. In a confined plasma where there is almost everywhere a non-zero pressure gradient there are well-defined surfaces of constant pressure. Magnetic field lines and lines of current flow lie on these surfaces, called magnetic surfaces or flux surfaces.

Equations (3.14) and (3.15) look quite simple, but construction of equilibria using them is rather complicated, particularly in full three-dimensional geometry. In experimental devices with some degree of symmetry the task is eased somewhat. We shall consider first the case of an axisymmetric toroidal plasma, a very important configuration since it includes the tokamak which is at present the most widely studied type of magnetic confinement device in nuclear fusion research. The relevant configuration is sketched in Figure 3.1, all quantities being independent of the coordinate ϕ.

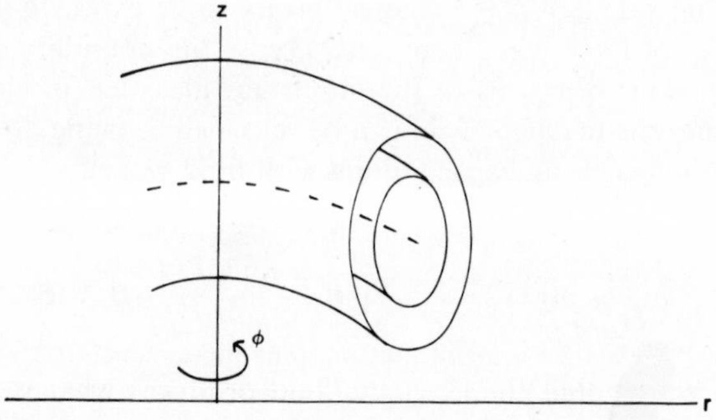

Figure 3.1 Magnetic surfaces in an axisymmetrical toroidal plasma. The dotted line, on which pressure is a maximum, is the magnetic axis.

If we introduce a coordinate ψ which is constant on a magnetic surface then

$$\boldsymbol{B}\cdot\boldsymbol{\nabla}\psi = 0,$$

or

$$B_r\frac{\partial\psi}{\partial r} + B_z\frac{\partial\psi}{\partial z} = 0. \tag{3.16}$$

Equation (3.16) together with the condition $\boldsymbol{\nabla}\cdot\boldsymbol{B} = 0$, can be satisfied if we set

$$B_r = -\frac{1}{r}\frac{\partial\psi}{\partial z}, \quad B_z = \frac{1}{r}\frac{\partial\psi}{\partial r}. \tag{3.17}$$

It may assist those with a knowledge of hydrodynamics to note that ψ is introduced in a manner similar to the stream function used in potential flow problems.

From the properties of the plasma discussed above it can be seen that the pressure is a function of ψ alone, i.e. $p = p(\psi)$. Also, any circle around the major axis can be labelled by its radius r and the magnetic surface on which it lies. Since current flow is along flux surfaces, the current flowing through the circle depends only on ψ and is independent of just where it is situated on the flux surface. We can, therefore, introduce a quantity $I(\psi)$, which is the current through a circle lying on the flux surface ψ, and note that from Ampère's law

$$B_\phi = \frac{\mu_0}{2\pi r}I(\psi). \tag{3.18}$$

With these relations the radial component of $\boldsymbol{J}\times\boldsymbol{B} = \boldsymbol{\nabla}p$ gives

$$r\frac{\partial}{\partial r}\frac{1}{r}\frac{\partial\psi}{\partial r} + \frac{\partial^2\psi}{\partial z^2} + \mu_0 r^2\frac{\partial p(\psi)}{\partial\psi} + \frac{\mu_0^2}{8\pi^2}\frac{\partial I^2(\psi)}{\partial\psi} = 0. \tag{3.19}$$

Here p and I can be chosen to be arbitrary functions of ψ, leaving a partial differential equation to be solved for $\psi(r, z)$ in order to construct a plasma equilibrium. At a perfectly conducting wall the boundary condition is that ψ is constant. In general (3.19) has to be solved numerically, though some analytic progress may be possible in special cases. One way of progressing is to treat a system whose minor radius, a, is small compared to its major radius, R, in which case an expansion in the quantity a/R, the inverse of the aspect ratio, may be useful. Here we shall confine ourselves to a discussion of some of the qualitative features of a solution relevant to a tokamak discharge.

First we see from (3.17) that if the plasma is contained in a toroidal discharge, with the pressure decreasing outwards from the centre to produce flux surfaces as in Figure 3.1, then B_r and B_z will in general be non-zero. A simple toroidal field with only B_ϕ non-zero is not sufficient to confine a plasma

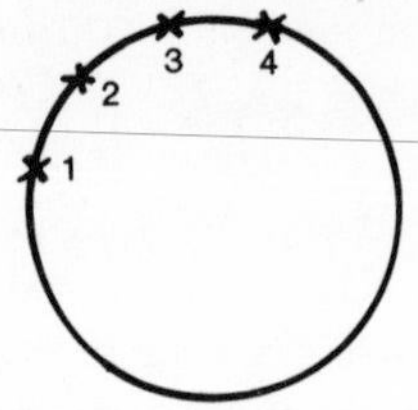

Figure 3.2 Successive intersections of a line of force with a minor cross-section.

in a toroidal system, a fact we have already noted in Chapter 2, from the point of view of orbit theory. In a tokamak the toroidal field B_ϕ is produced by means of external coils, while the poloidal component is generated by a current in the plasma. The major contribution to $I(\psi)$ is that produced by the external winding, since the main plasma current is in the toroidal direction, so that B_ϕ goes as $1/r$ across the minor axis of the torus. This toroidal field component is, in a tokamak, much stronger than the poloidal field component. In addition to the main external field produced by the toroidal windings it is also necessary to have a vertical field in the z-direction. This is because there is not in general a balance between the confining magnetic pressure due to the other field components and the tendency of the plasma ring to expand. Since there is a current around the plasma, a vertical field can produce a force on it directed towards the centre, and adjustments in the vertical field strength can be used to control the plasma position by changing its major radius. If a field line is followed around the torus, the successive intersections with a fixed minor cross-section are, as shown in Figure 3.2, displaced around the magnetic surface.

The average angle subtended at the magnetic axis by successive crossing points is called the rotational transform angle, and the safety factor, q, is 2π divided by this angle. This quantity is so called because it is related to the stability of the discharge and must not fall below unity.

Equation (3.19) just expresses the balance of plasma pressure and magnetic forces, the symmetry of the system having been used to describe the magnetic field in terms of a single parameter. The example of the axisymmetric system shows that, although the basic equations look simple, the actual calculation of equilibria in other than the simplest of geometries is a matter of some difficulty.

3.4 Stability of MHD equilibria

Having found an equilibrium configuration, the next question to ask is whether it is stable, that is, when it is perturbed slightly it simply oscillates about its equilibrium position, or whether the perturbation grows.

Ideally one would like to have a confinement system which was completely stable, but it is important to realize that this may not be possible and that the presence of weak, slowly growing instabilities may be acceptable provided that they do not lead to a complete break-up of the equilibrium configuration.

There are two ways which are commonly employed to investigate stability. The first and perhaps more obvious technique is to consider small perturbations about the equilibrium and to write each quantity as its equilibrium value plus a perturbation. The equations are then linearized, that is, products of perturbations are assumed negligible. In the resulting set of linear equations, the time variation is assumed to be as $e^{-i\omega t}$, so that we are left with differential equations in the spatial coordinates.

With suitable boundary conditions the problem becomes an eigenvalue problem for which solutions only exist for certain values of ω. If any of these values has a positive imaginary part, then the corresponding perturbation grows exponentially in time and the system is unstable. The other technique, which is particularly advantageous in complicated geometries, uses an energy principle. The basic idea is to consider the change in energy associated with a small perturbation. If there is a perturbation which lowers the potential energy of the system then this energy can be converted into kinetic energy and produce growth of the perturbation. This is just as in elementary mechanics where a system is unstable at a local maximum of the potential energy and stable when it is at rest at a minimum.

Taking the first approach and using a subscript 1 to denote the perturbations, we obtain the linearized equations

$$\frac{\partial \rho_1}{\partial t} + \boldsymbol{\nabla}\cdot(\rho \boldsymbol{V}_1) = 0 \tag{3.20}$$

$$\rho\frac{\partial \boldsymbol{v}_1}{\partial t} + \boldsymbol{\nabla} p_1 = \boldsymbol{J}\times\boldsymbol{B}_1 + \boldsymbol{J}_1\times\boldsymbol{B} \tag{3.21}$$

$$\frac{\partial p_1}{\partial t} + (\boldsymbol{v}_1\cdot\boldsymbol{\nabla})p + \gamma p\boldsymbol{\nabla}\cdot\boldsymbol{v}_1 = 0 \tag{3.22}$$

$$\frac{\partial \boldsymbol{B}_1}{\partial t} = \boldsymbol{\nabla}\times(\boldsymbol{v}_1\times\boldsymbol{B}). \tag{3.23}$$

The unsubscripted quantities are those of the equilibrium. If $\xi(r, t)$ is the displacement of the plasma from the equilibrium, then

$$\boldsymbol{v}_1 = \left(\frac{\partial}{\partial t} + \boldsymbol{v}_1\cdot\boldsymbol{\nabla}\right)\xi \approx \frac{\partial \xi}{\partial t}$$

and (3.23) gives

$$\boldsymbol{B}_1 = \boldsymbol{\nabla}\times(\boldsymbol{\xi}\times\boldsymbol{B}). \tag{3.24}$$

From (3.20) we get

$$\rho_1 = -\boldsymbol{\nabla}\cdot(\rho\boldsymbol{\xi}), \tag{3.25}$$

and from (3.22)

$$p_1 = -\boldsymbol{\xi}\cdot\boldsymbol{\nabla} p - \gamma p\boldsymbol{\nabla}\cdot\boldsymbol{\xi}, \tag{3.26}$$

and on substituting (3.25) and (3.26) into (3.20),

$$\rho\frac{\partial^2 \xi}{\partial t^2} = \nabla(\xi\cdot\nabla p + \gamma p\nabla\cdot\xi) + \frac{1}{\mu_0}(\nabla\times B)\times B_1 + \frac{1}{\mu_0}(\nabla\times B_1)\times B. \quad (3.27)$$

With perturbations going as $e^{-i\omega t}$, the left-hand side of (3.27) becomes $-\omega^2\rho\xi$, and (3.27) can be combined with (3.24) to give a single equation in ξ. With suitable boundary conditions this is the required eigenvalue problem which determines the possible values of ω^2.

We shall illustrate this by looking at the stability of a cylindrical plasma column, as illustrated in Figure 3.3. The equilibrium has uniform longitudinal fields B_i inside and B_e outside the sharp-edged plasma, with an azimuthal field

$$B_\theta = \frac{\mu_0 I}{2\pi r}$$

outside the column. This implies that there is a current I flowing along the plasma boundary in the z-direction. The displacement may be taken to be of the form $\xi(r)e^{ikz+im\theta}$, with m an integer, an arbitrary displacement being capable of Fourier decomposition into a superposition of such modes.

We shall consider incompressible modes, for which

$$\nabla\cdot\xi = 0. \quad (3.28)$$

Looking first at the interior of the plasma column we calculate from (3.24) and (3.28) that the perturbation to the magnetic field is just

$$B_1 = ikB_i\xi$$

while the pressure perturbation can be seen from (3.26) to be zero. Taking the divergence of (3.27) we have

$$B_i\cdot\nabla^2\xi = 0$$

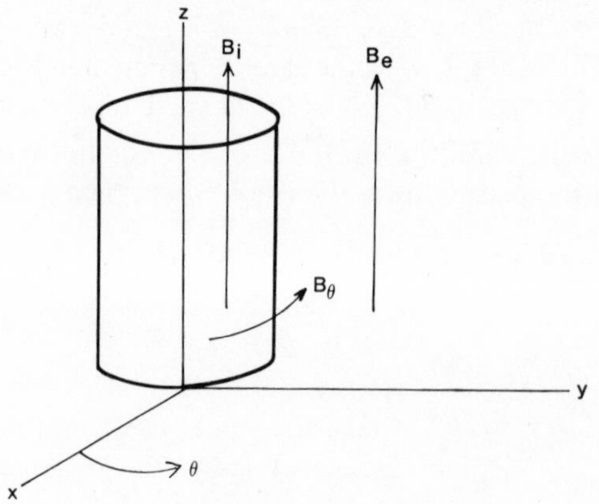

Figure 3.3 Cylindrical plasma equilibrium.

or

$$\nabla^2 \xi_z = 0,$$

which, with the assumed form of the perturbation gives

$$\left(\frac{d^2}{dr^2} + \frac{1}{r}\frac{d}{dr} - \left(k^2 + \frac{m^2}{r^2}\right)\right)\xi_z(r) = 0$$

The solution of this with the condition that there be no singularity at the origin is

$$\xi_z(r) = \xi_z(a)\frac{I_m(kr)}{I_m(ka)}, \tag{3.29}$$

with I_m the modified Bessel function. The radial component of $\boldsymbol{\xi}$ is then obtained from (3.27) and is

$$\xi_r(r) = \frac{ikB_i^2/\mu_0}{\omega^2\rho - \dfrac{k^2 B_i^2}{\mu_0}}\frac{d\xi_z(r)}{dr}. \tag{3.30}$$

In the vacuum region outside the plasma column $\nabla \times \boldsymbol{B}_1 = 0$ and so $\boldsymbol{B}_1$ can be obtained from a scalar potential ψ, with $\nabla^2\psi = 0$ and $\boldsymbol{B}_1 = -\nabla\psi$. With ψ of the form $\psi(r)e^{ikz+im\theta}$, $\psi(r)$ satisfies the same equation as ξ_z, the appropriate solution being now proportional to $K_m(kr)$, with K_m the modified Bessel function of the second kind, which is singular at the origin but has the property which we now require of going to zero at infinity. We shall take

$$\psi = C\frac{K_m(kr)}{K_m(ka)}e^{ikz+im\theta}, \tag{3.31}$$

with C a constant to be determined.

Having obtained the perturbations inside and outside the plasma column we now have to look at the conditions to be satisfied at the boundary between the two regions. The first such condition is that the pressure should remain continuous across the boundary. Since we have shown that there is no perturbation of the fluid pressure, the change in pressure inside the column is, to the first order, simply $B_i B_{1z}/\mu_0$, which at the boundary is

$$\frac{ikB_i^2}{\mu_0}\xi_z(a).$$

Similarly the change in pressure in the external region is

$$\frac{1}{\mu_0}(B_e B_{1z} + B_e B_{1\theta})$$

which, using (3.31), is found to take the value

$$\frac{1}{\mu_0}C\left(ikB_e + \frac{im}{a}B_\theta(a)\right)$$

at $r = a$, the exponential dependence being suppressed.

We must also take account of the fact that the B_θ part of the zero order field varies in the radial direction, so there is an additional perturbation to the external pressure given by

$$\xi_r(a)\frac{\partial}{\partial r}\left(\frac{B_\theta^2}{2\mu_0}\right)_{r=a}$$

which is

$$-\frac{B_\theta^2(a)}{\mu_0 a}\xi_r(a).$$

Our final boundary condition expressing continuity of pressure is thus

$$\frac{ikB_i^2}{\mu_0}\xi_z(a) = \frac{1}{\mu_0}C\left(ikB_e + \frac{im}{a}B_\theta(a)\right) - \frac{B_\theta^2(a)}{\mu_0 a}\xi_r(a). \tag{3.32}$$

Another boundary condition is the standard one of electromagnetic theory that the normal component of $\boldsymbol{B}$ be continuous across the boundary. The perturbation to this quantity is $\boldsymbol{B}_1\cdot\hat{\boldsymbol{n}} + \boldsymbol{B}\cdot\delta\hat{\boldsymbol{n}} + (\boldsymbol{\xi}\cdot\boldsymbol{\nabla})B_n$ where $\hat{\boldsymbol{n}}$ is the unit normal to the interface, $\delta\hat{\boldsymbol{n}}$ its change due to the motion of the boundary and B_n the normal component of the unperturbed field. The z-component of $\delta\hat{\boldsymbol{n}}$ is $-\partial\xi_r/\partial z$ and the θ component $-\partial\xi_r/\partial\theta$. Thus the normal component of $\boldsymbol{B}$ at the inner surface of the interface is

$$B_{1r}(a) - B_i ik\xi_r(a) = 0,$$

while that at the outer surface is

$$B_{1r}(a) - B_e\frac{\partial\xi_r}{\partial z} - B_\theta(a)\frac{\partial\xi_r}{\partial\theta},$$

which gives

$$Ck\frac{K_m'(ka)}{K_m(ka)} = i\left(kB_e + \frac{m}{a}B_\theta(a)\right)\xi_r(a). \tag{3.33}$$

Finally we use (3.29), (3.30), (3.32) and (3.33) to eliminate C and $\boldsymbol{\xi}(a)$, obtaining the dispersion relation

$$\frac{\omega^2}{k^2} = \frac{B_i^2}{\mu_0\rho} - \frac{\left(kB_e + \frac{m}{a}B_\theta(a)\right)^2}{\mu_0\rho k^2}\frac{I_m'(ka)K_m(ka)}{I_m(ka)K_m'(ka)} - \frac{B_\theta^2(a)}{\mu_0\rho}\frac{1}{(ka)}\frac{I_m'(ka)}{I_m(ka)}. \tag{3.34}$$

From this it can be seen that ω^2 is real, so that if $\omega^2 < 0$ instability occurs with a purely growing non-oscillatory mode. Since $K_m'(ka)$ is negative, the first two terms are positive, while the third is negative and if large enough can produce instability.

We shall look at a few special cases in order to illustrate the sort of

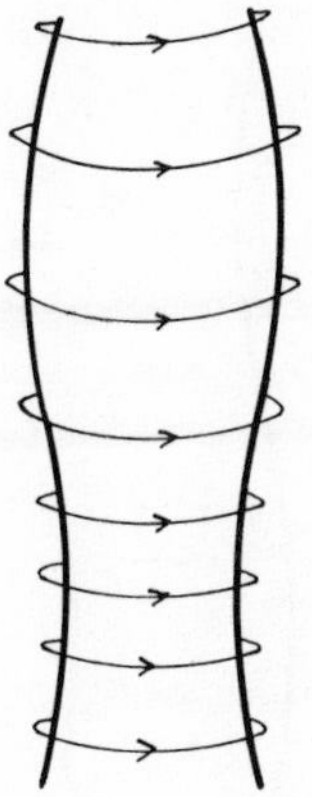

Figure 3.4 Plasma deformation in $m = 0$ or 'sausage' instability.

instability predicted by (3.34). First let us take the longitudinal field outside the plasma, i.e. B_e, to be zero, and look at $m = 0$ modes. Then (3.34) gives

$$\frac{\omega^2}{k^2} = \frac{B_i^2}{\mu_0 \rho} - \frac{B_\theta^2(a)}{\mu_0 \rho} \frac{I_0'(ka)}{kaI_0(ka)}.$$

Since $I_0'(x) < \frac{1}{2}xI_0(x)$ for all x, there is stability for all k if $B_i^2 > B_\theta^2(a)/2$. A perturbation with $m = 0$ has cylindrical symmetry and produces a configuration as shown in Figure 3.4. The physical reason for this so called sausage instability is quite easily seen. Where a narrower portion of the column develops, the current along the surface of the plasma is at a smaller radius, and so the field at the surface $I/2\pi\mu_0 r$ is greater. This gives an increased magnetic pressure which tends to increase the perturbation. An internal longitudinal field suppresses the instability because the compression of flux produced in a narrow part of the column produces an increased internal pressure. An external longitudinal field also stabilizes the column for a similar reason.

With $m = 1$ the plasma is deformed as shown in Figure 3.5, giving a 'kink mode'. Again the instability tends to occur if B_θ is large enough compared with the longitudinal field. The physical mechanism can be seen by referring to Figure 3.5, B_θ being enhanced on the side of the column on which increased pressure will cause the instability to grow. If a longitudinal field is present, either inside or outside, the tension forces produced by the bending of the field lines are such as to have a stabilizing effect.

Finally let us consider the case where the longitudinal fields are much greater than B_θ. This is important since it is the situation which obtains in a tokamak discharge, and the plasma cylinder might be expected to give some information about its stability, although any effects of the toroidal curvature of the column will be missing. Assuming also that $ka \ll 1$, and using the approximations $I_m(x) \approx (\frac{1}{2}x)^m/m!$ and $K_m(x) \approx \frac{1}{2}(\frac{1}{2}x)^{-m}$ for small x (and $m > 0$),

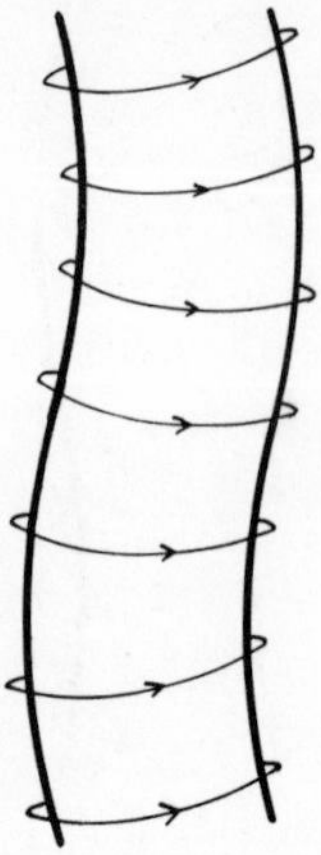

Figure 3.5 Plasma deformation with $m = 1$ or kink mode.

we have

$$\omega^2 \approx \frac{1}{\mu_0 \rho}\left[k^2 B_i^2 + \left(kB_e + \frac{m}{a}B_\theta(a)\right)^2 - \frac{m}{a^2}B_\theta^2(a)\right] \tag{3.35}$$

Here ω^2 is a quadratic in k, so that the minimum possible value of ω^2 as k varies is easily found to be

$$\omega_{\min}^2 = \frac{B_\theta^2(a)}{\mu_0 \rho a^2}\left(\frac{m^2 B_i^2}{B_e^2 + B_i^2} - m\right)$$

If the longitudinal fields are dominant, pressure balance in equilibrium requires that

$$p + \frac{B_i^2}{2\mu_0} = \frac{B_e^2}{2\mu_0}$$

If we define β to be the ratio of the plasma pressure to that of the external confining field, i.e.

$$\beta = \frac{p}{(B_e^2/2\mu_0)}$$

then

$$\omega_{\min}^2 = \frac{B_\theta^2(a)}{\mu_0 \rho a^2} m\left(m\frac{1-\beta}{2-\beta} - 1\right).$$

Tokamak equilibria exist only for low values of β, in which case only $m = 1$ and $m = 2$ modes can become unstable. If we look at the value of k for which $\omega_{\min}$ is obtained we find that $|ka| \ll 1$ if $B_\theta/B_e \ll 1$, so justifying our initial concentration on this long wavelength regime. From (3.35) it can be seen that the $m = 1$ mode is stable if $|ka| > |B_\theta(a)/B_e|$, so instability is only possible for k sufficiently small, that is for sufficiently long wavelength. In a plasma of finite length L, k cannot be less than $2\pi/L$, so we have the stability condition

$$\left|\frac{B_\theta}{B_z}\right| < \frac{2\pi a}{L},$$

known as the Kruskal–Shafranov condition. If we imagine the plasma column bent round to form a torus, then L becomes $2\pi R$ with R the major radius.

Our examination of the stability of a cylindrical column using a straightforward perturbation approach has shown that obtaining the dispersion relation is a rather complicated business, becoming, as may be imagined, much more difficult in more complex geometries. For this reason much work on MHD stability has approached the problem using an energy principle. A derivation of this principle starts from equation (3.27) which we can write as

$$\rho\frac{\partial^2 \xi}{\partial t^2} = \boldsymbol{F}(\xi), \tag{3.36}$$

where $\boldsymbol{F}$ is linear in ξ and can be interpreted as the force per unit volume acting on an element of the plasma when it is displaced by an amount ξ.

The change in potential energy during this displacement is given by

$$\delta W = -\frac{1}{2}\int_V \xi\cdot\boldsymbol{F}(\xi) \tag{3.37}$$

where the integral is throughout the plasma. If any possible displacement of the plasma is such as to make δW negative then the potential energy of the system is converted to kinetic energy and the plasma is unstable.

We shall omit the mathematical manipulations necessary to reduce (3.37) to a convenient form and simply state the form generally used and discuss its physical meaning. In a plasma with a boundary and a vacuum region outside, the energy integral can be split into three parts

$$\delta W = \delta W_p + \delta W_S + \delta W_V$$

giving the contributions of the plasma region V_p, the surface S and the external vacuum region V respectively. These three terms are

$$\begin{aligned}\delta W_p &= \frac{1}{2}\int_{V_p}\Big[\gamma p(\boldsymbol{\nabla}\cdot\xi)^2 + \frac{1}{\mu_0}(\boldsymbol{\nabla}\times(\xi\times\boldsymbol{B}))^2 + (\boldsymbol{\nabla}\cdot\xi)(\xi\cdot\boldsymbol{\nabla}p)\\ &\quad - \frac{1}{\mu_0}(\xi\times(\boldsymbol{\nabla}\times\boldsymbol{B}))\cdot(\boldsymbol{\nabla}\times(\xi\times\boldsymbol{B}))\Big]\,\mathrm{d}^3r\\ &= \frac{1}{2}\int_{V_p}\left[\frac{B_1^2}{\mu_0} - p_1(\boldsymbol{\nabla}\cdot\xi) - \xi(\boldsymbol{J}\times\boldsymbol{B}_1)\right]\mathrm{d}^3r\end{aligned} \tag{3.38}$$

$$\delta W_S = \frac{1}{2}\int_S(\xi\cdot\hat{\boldsymbol{n}})\frac{\partial}{\partial n}\left(\frac{B_\mathrm{e}^2}{2\mu_0} - \frac{B_\mathrm{i}^2}{2\mu_0} - p\right)\mathrm{d}S \tag{3.39}$$

$$\delta W_V = \frac{1}{2\mu_0}\int_{V_e} \frac{B_1^2}{2\mu_0}\mathrm{d}^3r. \tag{3.40}$$

The physics underlying δW_p is best seen from the second expression given in (3.38). The first term gives the contribution from the magnetic field perturbation. In the second, $\nabla\cdot\xi$ is related to the change in volume of a plasma element, and so this term gives the potential energy given up by expansion of the plasma. The third term is the work done by the magnetic force generated by the equilibrium current. In the surface integral the term $(B_e^2/2\mu_0) - (B_i^2/2\mu_0) - p$ is the pressure difference across the boundary, which vanishes in equilibrium when $\xi = 0$. A displacement $\xi_n = (\xi\cdot\hat{n})$ of the boundary ($\hat{n}$ being the outward normal) gives rise to a force per unit area

$$\xi_n \frac{\partial}{\partial n}\left(\frac{B_e^2}{2\mu_0} - \frac{B_i^2}{2\mu_0} - p\right),$$

and the work done per unit area as the displacement increases from zero is

$$\int_0^{\xi_n} \xi_n' \frac{\partial}{\partial n}\left(\frac{B_e^2}{2\mu_0} - \frac{B_i^2}{2\mu_0} - p\right)\mathrm{d}\xi_n' = \tfrac{1}{2}\xi_n^2 \frac{\partial}{\partial n}\left(\frac{B_e^2}{2\mu_0} - \frac{B_i^2}{2\mu_0} - p\right).$$

Thus (3.39) is the total work done in perturbing the boundary. The final term (3.40) is the contribution from the magnetic field outside the plasma. Note that in (3.38) the term in $\nabla\cdot\xi$ is stabilizing, so it is useful to consider incompressible perturbations as we did for the cylindrical plasma, these being the most unstable in general.

The second term in (3.38) vanishes if ξ is perpendicular to $\boldsymbol{B}$ and is otherwise stabilizing. This means that the most unstable modes tend to have perturbations aligned along magnetic field lines. In a toroidal system like a tokamak, such perturbations can occur on flux surfaces where the magnetic field lines meet up with themselves after a few turns around the torus. Perturbations following such lines will have the required periodicity around the toroidal direction. Instabilities are thus localized around those flux surfaces where the safety factor q, defined in section 3.2, is a small integer. The Kruskal–Shafranov condition, which we obtained earlier for a sharp-edged plasma, says that $q > 1$ at the boundary. For a diffuse plasma the condition is that $q > 1$ everywhere inside the plasma, so that there is no flux surface on which $m = 1$ perturbations can follow field lines around the torus.

A somewhat simplified use of energy arguments in the investigation of plasma stability can be seen in the following treatment of a type of instability known as an interchange instability. Let us consider a perturbation elongated along the magnetic field, taking in particular a perturbation which interchanges the plasma in two flux tubes containing equal magnetic fluxes, as illustrated in Figure 3.6. This leaves the magnetic energy unchanged, but may lead to a change in the internal energy of the plasma, which is proportional to pV. If p_1 and V_1 are the pressure and volume of flux tube 1, then when it is

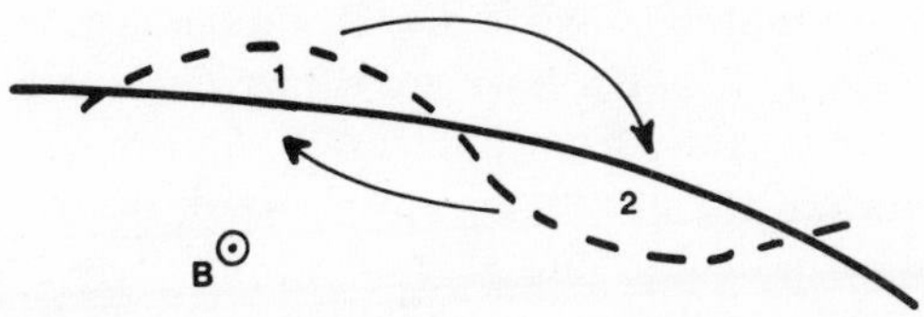

Figure 3.6 Cross-section, perpendicular to the magnetic field of the plasma in the interchange instability.

interchanged with flux tube 2 its volume changes to V_2 and, since the change is adiabatic, the change in its energy depends on

$$p_1\left(\frac{V_2}{V_1}\right)^{\gamma} V_2 - p_1 V_1,$$

the final value of pV minus the initial value. Similarly the change of energy of the plasma initially in position 2 is proportional to

$$p_2\left(\frac{V_1}{V_2}\right)^{c} V_1 - p_2 V_2.$$

If $p_2 = p_1 + \delta p$, $V_2 = V_1 + \delta V$, the sum of these two quantities is, to lowest order

$$\delta p \delta V + \gamma p_1 \frac{(\delta V)^2}{V_1}.$$

This will certainly be positive if $\delta p \delta V > 0$, so this gives a sufficient criterion for stability against this type of perturbation.

Now, the pressure generally decreases outwards from the centre of the plasma so that for stability we would like the volume occupied by a flux tube to decrease away from the centre. This volume is

$$\int S \, \mathrm{d}l,$$

l being measured along the length of the tube and S being its cross section. The flux is $\Phi = SB =$ constant, so the decrease in volume away from the centre may be expressed as requiring that

$$\int \frac{\mathrm{d}l}{B}$$

should decrease away from the centre. This condition can be interpreted as saying that B, suitably averaged along field lines, should increase away from the centre of the plasma, or that the plasma should occupy a region of average minimum B.

In a low-β plasma ($\beta =$ plasma pressure/magnetic pressure) the field is mainly due to the external coils, and is close to a vacuum field. In this case $\nabla \times \boldsymbol{B} \approx 0$ which implies that if field lines are concave towards the plasma the field strength decreases away from the plasma. Thus we may expect that a configuration in which field lines are concave towards the plasma, for example a simple mirror field, would be unstable to perturbations producing ripples aligned along the magnetic field, an instability known as the flute instability. In a machine like a tokamak the field curvature is favourable on the inner side of the torus and unfavourable on the outside. The average minimum B condition gives a condition for stability against a particular type of instability but does not, of course, rule out instability arising from different sorts of perturbations. Considerable attention is presently being given to instabilities known as ballooning modes, in which the perturbation is localized in regions of unfavourable field line curvature. Simple interchange instabilities can be inhibited by having a shear in the magnetic field making interchange of flux tubes as described above impossible. Such more complicated geometries can be investigated using the energy principle, but it should be realized that MHD stability is still very much a subject of active investigation.

In the above we have considered only instabilities in plasmas with zero electric resistivity. It is possible sometimes for a plasma with non-zero resistivity to be unstable even though the same configuration is stable in an ideal plasma. The reason why this should occur is connected with the fact that in an ideal plasma magnetic field lines are frozen into the fluid, so that the topology of the field lines cannot change. A finite conductivity relaxes this condition and may make it possible for the field to evolve and give up energy in a way which is forbidden in an ideal plasma. One of the most familiar examples is the tearing mode, in which a current sheet in the plasma produces a magnetic field as shown in Figure 3.7(*a*). The current flows perpendicular to the plane of the diagram, intersecting it along the dotted line. As is well known, parallel current filaments attract each other, so the system may move to a state of lower magnetic field energy if the current tends to split up into filaments, producing the field configuration of Figure 3.7(*b*). However this requires a breaking and rejoining of the field lines (tearing), and is not possible if the fields are frozen into the plasma. The tearing mode involves perturbations elongated along the field lines, and confined to a rather narrow layer.

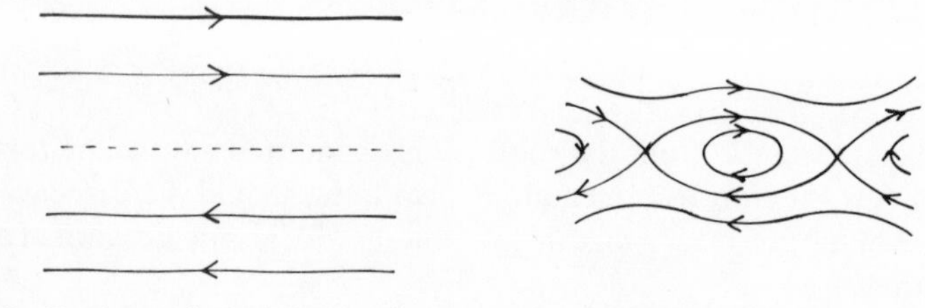

Figure 3.7 Tearing mode: (*a*), initial configuration, (*b*), magnetic field resulting from instability.

With this we shall leave the subject of MHD stability. This brief account makes no attempt to be exhaustive, but is intended to give some idea of the techniques used and of some representative results. In nuclear fusion research and in astrophysical applications of magnetohydrodynamics, stability is a topic of great importance and a great deal of effort has been, and is being, expended on it.

3.5 MHD waves

We now turn to the wave motions which are possible in a uniform conducting fluid, which we shall again assume to be perfectly conducting and to behave adiabatically. The general method used to investigate wave motions is to assume small perturbations about the equilibrium and to linearize the equations, just as was done in the last section to investigate stability. However, the emphasis now is not on whether or not the perturbations grow, since the first step at least is to consider waves in a uniform stable system. Rather the object is to investigate the propagation characteristics of small-amplitude waves.

Our starting point is just the set of linearized equations of motion (3.20)–(3.23), assuming now that in equilibrium p and B are uniform and the current is zero. Under these conditions (3.27) reduces to

$$\rho\frac{\partial^2 \xi}{\partial t^2} = \nabla(\gamma p \nabla\cdot\xi) + \frac{1}{\mu_0}(\nabla \times \boldsymbol{B}_1) \times \boldsymbol{B}, \tag{3.41}$$

with, as before

$$\boldsymbol{B}_1 = \nabla \times (\xi \times \boldsymbol{B}). \tag{3.42}$$

In a uniform system we can look for a plane wave solution, with all variables going as $e^{i\boldsymbol{k}\cdot\boldsymbol{r} - i\omega t}$, an arbitrary disturbance being a superposition of these. The gradient operator is then just replaced with $i\boldsymbol{k}$ and (3.41) and (3.42) become

$$-\omega^2\rho\xi = -\boldsymbol{k}\gamma p(\boldsymbol{k}\cdot\xi) + \frac{1}{\mu_0}(i\boldsymbol{k} \times \boldsymbol{B}_1) \times \boldsymbol{B} \tag{3.43}$$

and

$$\boldsymbol{B}_1 = i\boldsymbol{k} \times (\xi \times \boldsymbol{B}). \tag{3.44}$$

Letting $\boldsymbol{B}$ be along the z-axis and the angle between $\boldsymbol{k}$ and $\boldsymbol{B}$ be θ we obtain, by taking the scalar product of (3.43) with $\boldsymbol{k}$, $\hat{z}$ and $\boldsymbol{k} \times \hat{z}$ respectively, the three equations

$$\omega^2(\xi\cdot\boldsymbol{k}) = c_S^2 k^2(\xi\cdot\boldsymbol{k}) - c_A^2 \cos\theta\, \xi_z k^3 + c_A^2 k^2(\xi\cdot\boldsymbol{k}) \tag{3.45}$$

$$\omega^2 \xi_z = c_S^2 k \cos\theta(\xi\cdot\boldsymbol{k}) \tag{3.46}$$

$$(\omega^2 - c_A^2 k^2 \cos^2\theta)(\boldsymbol{k} \times \hat{z})\cdot\xi = 0, \tag{3.47}$$

where $c_S^2 = \gamma p/\rho$ and $c_A^2 = B^2/\mu_0\rho$. From these three equations, which are independent except in the special case when $\boldsymbol{k}$ is parallel to $\boldsymbol{B}$, we can deduce the dispersion relation, that is the relation between ω and $\boldsymbol{k}$, for the various possible wave modes.

First (3.47) shows that there is a mode with

$$\omega^2 = c_A^2 k^2 \cos^2\theta = c_A^2 k_z^2. \tag{3.48}$$

This is the Alfvén wave and c_A is the Alfvén speed. It is a mode which has $\boldsymbol{\xi}$ along the direction of $\boldsymbol{k} \times \hat{z}$, that is the plasma displacements are perpendicular to both the magnetic field and the wave vector. Since ω depends only on k_z the group velocity is along the magnetic field direction and these waves carry energy only in this direction. Equations (3.45) and (3.46) are coupled together, and by eliminating k_z and $(\boldsymbol{k}\cdot\boldsymbol{\xi})$ between them we can arrive at the dispersion relation

$$\omega^4 - k^2\omega^2(c_S^2 + c_A^2) + k^4 c_S^2 c_A^2 \cos^2\theta = 0,$$

giving

$$\frac{\omega^2}{k^2} = \tfrac{1}{2}\{c_S^2 + c_A^2 \pm [(c_S^2 + c_A^2)^2 - 4c_S^2 c_A^2 \cos^2\theta]^{1/2}\} \tag{3.49}$$

There are therefore two wave modes which in general will have both ξ_z and $\boldsymbol{k}\cdot\boldsymbol{\xi}$ non-zero. The normal sound speed of the unmagnetized gas is c_S and these modes are known as the fast and slow magnetosonic waves, the fast corresponding to the + sign and the slow to the − sign. If $\boldsymbol{k}$ is parallel to $\boldsymbol{B}$ the fast magnetosonic wave speed becomes c_S. In this case it has only the z-component of $\boldsymbol{\xi}$ non-zero and so the plasma displacements are along the magnetic field which has no effect on them. Perpendicular to the field the fast magnetosonic wave speed is $(c_S^2 + c_A^2)$. In this limit it is again a longitudinal wave, but now the magnetic field is compressed during the plasma motion and its contribution to the pressure adds to the gas pressure. The slow magnetosonic wave speed reduces to c_A for parallel propagation, and in this limit is the same as the Alfvén wave, being purely transverse. The Alfvén wave has $\boldsymbol{k}\cdot\boldsymbol{\xi} = 0$, that is $\nabla\cdot\boldsymbol{\xi} = 0$, so that it involves incompressible motion of the plasma. This is why its propagation is independent of the plasma pressure. The other two modes have non-zero $\boldsymbol{k}\cdot\boldsymbol{\xi}$, except for the slow mode with $\boldsymbol{k}$ along $\boldsymbol{B}$, and so bring in the sound speed. In general the group velocity of the magnetosonic waves is neither along the field nor along the direction of the phase velocity. However, in the limit where the plasma pressure becomes negligible the slow wave disappears while the fast wave propagates in a spherically symmetric fashion with $\omega^2 = k^2 c_A^2$. In this limit it is often known as the fast or compressional Alfvén wave.

3.6 Coulomb collisions

So far in this chapter we have described the plasma using fluid equations, whose use is most easily justified in a plasma which is collision-dominated. Thus they apply to slow, large length-scale phenomena as explained in

Chapter 1. At the other end of the scale are fast phenomena like high-frequency waves and instabilities, where collisions play no significant role and which will be discussed in later chapters. Between these two extremes are cases where the nature of the collisions is important. In this section we shall discuss some of the basic properties of Coulomb collisions, then in subsequent sections sketch out something of the kinetic theory of collisional plasmas and how it can be used to derive the fluid equations.

We begin by considering a collision between two particles of masses m_1 and m_2 and charges q_1 and q_2, interacting through the Coulomb force. The behaviour of particles interacting through an inverse square law is a well-known problem in classical mechanics, with applications to astronomy and to Rutherford scattering of charged particles from nuclei. We shall simply state the results we require.

The collision is most easily treated in a frame of reference in which the centre of mass is at rest, in which case the particle orbits are as in Figure 3.8. The particles are deflected through the same angle, while the energy of each remains constant. The relation between the impact parameter b and the angle of deflection is

$$\cot\frac{\psi}{2} = \frac{4\pi\varepsilon_0 \mu b u^2}{q_1 q_2}, \tag{3.50}$$

where μ is the reduced mass $m_1 m_2/(m_1 + m_2)$ and u is the relative velocity of the particles before the collision. For small-angle scattering we may approximate (3.50) by

$$\frac{\psi}{2} \approx \frac{q_1 q_2}{2\pi\varepsilon_0 \mu b u^2}. \tag{3.51}$$

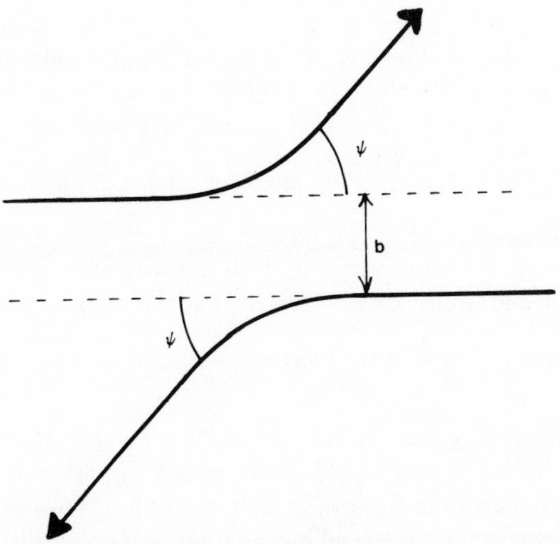

Figure 3.8 Particle orbits (for a repulsive force) in the centre of mass frame.

Now, let us consider the change in energy of particle 1 say, in the laboratory frame. If the velocities before and after the collision are $\boldsymbol{u}_1$ and $\boldsymbol{u}_1 + \delta\boldsymbol{u}_1$ in the centre of mass frame, then in the laboratory frame they are

$$\boldsymbol{v}_1 = \boldsymbol{u}_1 + \boldsymbol{V}$$

and

$$\boldsymbol{v}_1 + \delta\boldsymbol{v}_1 = \boldsymbol{u}_1 + \delta\boldsymbol{u}_1 + \boldsymbol{V},$$

where $\boldsymbol{V}$ is the centre of mass velocity. Since the kinetic energy of each particle in the centre of mass frame is the same before and after the collision, $2\delta\boldsymbol{u}_1 \cdot \boldsymbol{u}_1 + \delta u_1^2 = 0$, and it follows that the change of energy in the laboratory frame is

$$\delta\varepsilon_1 = 2m_1 \boldsymbol{V} \cdot \delta\boldsymbol{u}_1. \tag{3.52}$$

Now the component of $\delta\boldsymbol{u}_1$ along the direction of approach of the particles is $u_1(1 - \cos\psi) \approx u_1(\psi^2/2)$ for small ψ, while that at right angles to this direction is $u_1 \sin\psi$. If we consider one particle of species 1, a 'test particle', and imagine it to be colliding with an ensemble of particles of type 2 of density n_2, the number of collisions per unit time with impact parameters between b and $b + \mathrm{d}b$ is

$$n_2 u 2\pi b \, \mathrm{d}b.$$

The rate of change of energy due to the collisions is found from (3.52) by averaging over all possible impact parameters. The contribution from the change in $\boldsymbol{u}_1$ perpendicular to the line of approach, being proportional to $\sin\psi$, averages to zero, but from the component along the line of approach we obtain

$$\frac{\mathrm{d}\varepsilon_1}{\mathrm{d}t} = -\int m_1 \boldsymbol{u}_1 \cdot \boldsymbol{V} \frac{\psi^2}{2} n_2 u 2\pi b \, \mathrm{d}b,$$

which on expressing ψ in terms of b using (3.51) and noting that $m_1\boldsymbol{u}_1 = \mu\boldsymbol{u}$ gives

$$\frac{\mathrm{d}\varepsilon_1}{\mathrm{d}t} = -\frac{q_1^2 q_2^2 n_2 \boldsymbol{V} \cdot \boldsymbol{u}}{4\pi\varepsilon_0^2 \mu u^3} \int \frac{\mathrm{d}b}{b}. \tag{3.53}$$

In (3.53) the range of integration in b might be thought to be from zero to infinity, but this gives a divergence at both ends of the range. We must choose maximum and minimum values of b, so that the integral becomes

$$\log \frac{b_{\max}}{b_{\min}} = \log \Lambda.$$

Taking the maximum first we remember that, as shown in Chapter 1, the interaction between two particles in a plasma is affected by the Debye shielding. For distances less than the Debye length it is roughly the Coulomb potential but then begins to fall off more rapidly at greater distances. Taking a rough approximation to be a Coulomb potential cut off at the Debye length, we choose $b_{max} = \lambda_D$. At the lower end we note that we have assumed, in using (3.51), small-angle scattering, which is not valid for small b. We therefore cut off the integral at a value b_{min} corresponding to the distance at which the potential energy of the interaction is equal to the average particle kinetic energy in the centre of mass frame, i.e. $b_{min} = q_1 q_2/(4\pi\varepsilon_0 \mu \bar{u}^2)$, with $\bar{u}^2$ found from the average over the particle velocity distributions. Alternatively, if this distance turns out to be less than the de Broglie wavelength of the particles, quantum effects are important and this wavelength may be taken as b_{min}. This definition is appropriate when the electron temperature is greater than about 100 eV. It is fortunate that the dependence on b_{max} and b_{min} is logarithmic and that for most laboratory plasmas it takes values between 10 and 20. Thus the exact choice of b_{max} and b_{min} is not of great importance. The fact that b_{max} is many orders of magnitude greater than b_{min} shows that, as we have claimed, most scatterings are through small angles.

The rate of change of energy

$$\frac{d\varepsilon_1}{dt} = -\frac{q_1^2 q_2^2 n_2 \boldsymbol{V}\cdot\boldsymbol{u}}{4\pi\varepsilon_0^2 \mu u^3} \log \Lambda \tag{3.54}$$

should now be averaged over the velocities of the particles 2, so that the final expression for the rate of change of energy of the test particle 1 due to collisions with the distribution of particles of type 2 is

$$\frac{d\varepsilon_1}{dt} = -\frac{q_1^2 q_2^2 n_z}{4\pi\varepsilon_0^2 \mu} \log \Lambda \int \frac{\boldsymbol{V}\cdot\boldsymbol{u}}{u^3} f_2(\boldsymbol{v}_2)\, d^3\boldsymbol{v}_2, \tag{3.55}$$

where $\boldsymbol{V}$ and $\boldsymbol{u}$, of course, depend on $\boldsymbol{v}_2$ and on the test particle velocity. The total rate of change of energy can then be obtained by summing (3.55) over all the particle species in the plasma with which the test particle is interacting.

Using (3.55) we may obtain some estimate of the characteristic time over which a distribution somewhat different from a Maxwellian relaxes back to a Maxwellian. Let us consider first the background particles to be a distribution of electrons with temperature T_e and consider the relaxation of a distribution of test particles with a slightly different temperature T_e'. For electron–electron collisions $\mu = \frac{1}{2} m_e$, $\boldsymbol{V} = \frac{1}{2}(\boldsymbol{v}_1 + \boldsymbol{v}_2)$ and $\boldsymbol{u} = \boldsymbol{v}_1 - \boldsymbol{v}_2$, so that (3.54) is

$$\frac{d\varepsilon_1}{dt} = -\frac{e^4 n_e}{4\pi\varepsilon_0^2 m_e} \frac{\boldsymbol{v}_1^2 - \boldsymbol{v}_2^2}{|\boldsymbol{v}_1 - \boldsymbol{v}_2|^3} \log \Lambda .$$

We wish to average this over both background and test particle distributions, and to simplify the calculation we shall replace $|\boldsymbol{v}_1 - \boldsymbol{v}_2|$ in the denominator by a typical value which we assume to be the thermal velocity $(3\kappa T_e/m_e)^{1/2}$. Then we obtain

$$\frac{dT_e'}{dt} \approx -\frac{e^4 n_e}{6\pi\varepsilon_0^2 m_e^{1/2}} \frac{T_e' - T_e}{3^{1/2}(\kappa T_e)^{3/2}} \log \Lambda,$$

from which we may estimate the time over which the temperature of the test particles comes into equilibrium with that of the background as a result of electron–electron collisions to be

$$\tau_{ee} \approx \frac{3^{1/2} 6\pi\varepsilon_0^2 (\kappa T_e)^{3/2} m_e^{1/2}}{n_e e^4 \log \Lambda} \tag{3.56}$$

In a similar way the time associated with relaxation of ions to equilibrium as a result of ion–ion collisions is

$$\tau_{ii} \approx \frac{3^{1/2} 6\pi\varepsilon_0^2 (\kappa T_e)^{3/2} m_e^{1/2}}{n_i Z^4 e^4 \log \Lambda}, \tag{3.57}$$

where Z is the ion charge number.

Finally let us consider the equilibration of electrons and ions at slightly different temperatures. For electron–ion collisions $\mu \approx m_e$, $\boldsymbol{V} \approx (m_e \boldsymbol{v}_e + m_i \boldsymbol{v}_i)/m_i$, $\boldsymbol{u} = \boldsymbol{v}_e - \boldsymbol{v}_i$ and so

$$\boldsymbol{V}\cdot\boldsymbol{u} = \frac{m_e}{m_i} v_e^2 - v_i^2 + \boldsymbol{v}_e\cdot\boldsymbol{v}_i \left(\frac{m_e}{m_i} + 1 \right).$$

Averaged over isotropic distributions of electrons and ions we would expect the contribution from $\boldsymbol{v}_e\cdot\boldsymbol{v}_i$ to vanish, and the average value of $\boldsymbol{V}\cdot\boldsymbol{u}$ to be $(2/m_i)(\bar{\varepsilon}_e - \bar{\varepsilon}_i)$. Typically the relative velocity u will again be of the order of the electron thermal velocity, so that from (3.54) we can estimate that

$$\frac{\mathrm{d}T_e}{\mathrm{d}t} \approx -\frac{Z^2 e^4 n_i}{4\pi\varepsilon_0^2 m_e \left(\dfrac{3\kappa T_e}{m_e} \right)^{3/2}} \frac{2}{m_i} (T_e - T_i) \log \Lambda,$$

giving us a characteristic time for electron–ion temperature equilibration of

$$\tau_{ei} \approx \frac{3^{1/2} 6\pi\varepsilon_0^2 (\kappa T_e)^{3/2} m_i}{Z^2 n_i e^4 m_e^{1/2} \log \Lambda}. \tag{3.58}$$

Comparing (3.56)–(3.58) shows that the electrons come to thermal equilibrium in the shortest time, with the ions taking a time of order $(1/Z^3)(m_i/m_e)^{1/2}$ longer to establish equilibrium (note that $n_e = Z n_i$). Finally equalization of ion and electron temperatures takes a longer time still, of the order of $(1/Z)(m_i/m_e)$ times that required for electron thermalization. More careful evaluation of these relaxation times may change the numerical constants which appear in (3.56)–(3.58), but the essential dependences on the plasma parameters are correct. The long time required for equilibrium to be established between the ion and electron temperatures has the result that in many plasma experiments different electron and ion temperatures can exist, with each species having a distribution close to Maxwellian.

Somewhat similar considerations may be used to make an estimate of plasma resistivity. If η is the resistivity then $\boldsymbol{E} = \eta \boldsymbol{J}$ and the rate of increase of electron momentum per unit volume due to the field is

$$-en_e\boldsymbol{E} = -en_e\eta j = e^2 n_e \eta \boldsymbol{v}_e.$$

This must be balanced by electron–ion collisions since electron–electron collisions have no effect on the total electron momentum. The change in momentum of an electron in a collision with an ion is

$$m_e\boldsymbol{u}(1-\cos\psi) \approx m_e\boldsymbol{u}\frac{\psi^2}{2},$$

taking the component along the direction of approach in the collision, since other components average to zero. Averaging this over impact parameters gives

$$\frac{\mathrm{d}\boldsymbol{p}_e}{\mathrm{d}t} = -\frac{Z^2e^4n_i}{4\pi\varepsilon_0^2\mu^2u^3}m_e\boldsymbol{u}\log\Lambda. \tag{3.59}$$

By inserting for u in the denominator the average electron velocity $(3\kappa T_e/m_e)$ we can, as before, estimate the average rate of change of momentum due to electron–ion collisions to be

$$\frac{\mathrm{d}\boldsymbol{p}_e}{\mathrm{d}t} \approx -\frac{Ze^4n_e}{3^{1/2}12\pi\varepsilon_0^2(\kappa T_e)^{3/2}}m_e^{1/2}\boldsymbol{v}_e\log\Lambda.$$

Balancing this against the momentum change produced by the field gives

$$\eta = \frac{Ze^2m_e^{1/2}\log\Lambda}{3^{1/2}12\pi\varepsilon_0^2(\kappa T_e)^{3/2}}. \tag{3.60}$$

Again, a more exact averaging over the velocity distribution may change the numerical factors in (3.60), but the basic form is correct. A feature of this result is that resistivity decreases with temperature, but is independent of density. This has important consequences as regards the heating of plasmas in tokamak discharges, making resistive heating ineffective when the temperature reaches 2–3 keV and necessitating the use of some auxiliary heating scheme if the temperatures needed for nuclear fusion are to be reached.

A consequence of (3.59) and of the corresponding expression for slowing down of a test electron owing to electron–electron collisions, is that the drag on an electron decreases with its velocity. If an electric field is applied to a plasma, the drag on a sufficiently fast electron may be less than the electric field force eE, so that the electron is continuously accelerated. Electrons behaving in this way are called runaway electrons.

3.7 The Fokker–Plank equation

If a more exact theory of the behaviour of non-equilibrium particle distribution functions under the effect of collisions is required then we must obtain a kinetic equation describing their evolution in time. The kinetic equation which is most often used in practical applications is generally referred to as the Fokker–Plank equation, owing to its resemblance to the original equation of this name which was proposed to describe the behaviour of particles undergoing Brownian motion.

The Fokker–Plank equation describes the behaviour of a collection of particles undergoing a series of small-angle scatterings and its general form may be obtained as follows. Assuming for the present that the plasma is spatially homogeneous we define the distribution function $f_s(\boldsymbol{v}, t)$ of the species s to be such that the number of particles of this species per unit volume in an element d^3v of velocity space at time t is $f_s(\boldsymbol{v}, t)\,d^3v$. The total density n_s is then the integral of f_s over all velocities. The distribution at time $t + \Delta t$ may be expressed in terms of that at time t through the equation

$$f(\boldsymbol{v}, t + \Delta t) = \int f(\boldsymbol{v} - \Delta\boldsymbol{v}, t) P(\boldsymbol{v} - \Delta\boldsymbol{v}, \Delta\boldsymbol{v})\, d^3\Delta v, \tag{3.61}$$

where $P(\boldsymbol{v}, \Delta\boldsymbol{v})$ is the probability that a particle which has velocity $\boldsymbol{v}$ at time t receives a velocity increment $\Delta\boldsymbol{v}$ in the time interval between t and $t + \Delta t$. Equation (3.61) just says that the density of particles at a point in velocity space at time $t + \Delta t$ is given by an integral over densities in neighbouring regions of velocity space at time t, multiplied by the probability that particles in these regions move to the point of interest.

If we assume now that the functions appearing in (3.61) are smooth functions of t and $\boldsymbol{v}$ than we may make a Taylor expansion so that (3.61) becomes

$$f(\boldsymbol{v}, t) + \frac{\partial f(\boldsymbol{v}, t)}{\partial t}\Delta t \approx \int \left\{ f(\boldsymbol{v}, t) P(\boldsymbol{v}, \Delta\boldsymbol{v}) - \frac{\partial (fP)}{\partial v_i}\Delta v_i + \frac{1}{2}\frac{\partial^2 (fP)}{\partial v_i \partial v_j}\Delta v_i \Delta v_j \right\} d^3\Delta v, \tag{3.62}$$

where we keep terms to first order in Δt and second order in $\Delta\boldsymbol{v}$. In (3.62), tensor notation with $\boldsymbol{v} = (v_1, v_2, v_3)$ and summation over repeated indices is to be understood in the right-hand side. Such an expansion is valid because most plasma collisions are small-angle scatterings, so that particles change continuously in velocity rather than undergoing abrupt changes as would happen with the billiard-ball-like collisions of neutral gas molecules.

From the definition of P as a probability we can say that

$$\int P(\boldsymbol{v}, \Delta\boldsymbol{v})\, d^3\Delta v = 1.$$

Also,

$$\int P(\boldsymbol{v}, \Delta\boldsymbol{v}) \Delta v_i\, d^3\Delta v$$

is the average value of Δv_i, while

$$\int P(\boldsymbol{v}, \Delta \boldsymbol{v}) \Delta v_j \, d^3 \Delta v$$

is the average of $\Delta v_i \Delta v_j$. Using angular brackets to denote these averages we may write (3.62) as

$$\frac{\partial f(\boldsymbol{v}, t)}{\partial t} \Delta t = -\frac{\partial}{\partial v_i}(\langle \Delta v_i \rangle f) + \frac{1}{2} \frac{\partial^2}{\partial v_i \partial v_j}(\langle \Delta v_i \Delta v_j \rangle f).$$

As we shall see shortly both of the averages in this equation are proportional to Δt, so that

$$\frac{\partial f}{\partial t} = -\frac{\partial}{\partial v_i}(A_i f) + \frac{1}{2} \frac{\partial}{\partial v_i}\left(D_{ij} \frac{\partial f}{\partial v_j}\right), \tag{3.63}$$

where

$$A_i = \frac{\langle \Delta v_i \rangle}{\Delta t} - \frac{1}{2} \frac{\partial}{\partial v_j} \frac{\langle \Delta v_i \Delta v_j \rangle}{\Delta t}; \quad D_{ij} = \frac{\langle \Delta v_i \Delta v_j \rangle}{\Delta t}.$$

Equation (3.63) is the Fokker–Plank equation. The first term on the right-hand side comes from the slowing down of the particle owing to collisions, and is referred to as the friction term, while the second term represents diffusion of particles in velocity space. In a plasma the equilibrium Maxwellian distribution results from a balance between the friction term, tending to concentrate the particles around the mean flow velocity of the plasma, and the diffusion term which tends to spread particles in velocity space.

We now turn to the question of calculating the coefficients A_i and D_{ij}. From considerations like those leading to (3.59) we may calculate $\langle \Delta v_i \rangle / \Delta t$ due to collisions of a particle of species s with all the other particles in the plasma to be

$$\frac{\langle \Delta v_i \rangle}{\Delta t} = -\frac{q_s^2}{4\pi\varepsilon_0^2 m_s} \sum_{s'} \frac{q_{s'}^2 \log \Lambda}{\mu_{ss'}} \int \frac{u_i}{u^3} f_{s'}(\boldsymbol{v}') \, d^3 v' \tag{3.64}$$

with $\boldsymbol{u} = \boldsymbol{v} - \boldsymbol{v}'$ and $\mu_{ss'}$ the reduced mass corresponding to m_s and $m_{s'}$.

To calculate $\langle \Delta v_i \Delta v_j \rangle$ we suppose that the particle velocity is along the z-axis and consider the change in velocity in the x-direction. Then the change in v_x due to the collision is

$$\delta v_x = \delta v_\perp \cos \theta,$$

where θ is the angle shown in Figure 3.9 and $\delta v_\perp$ is the change in total speed perpendicular to the line of approach of the particles. If we consider N such collisions, each scattering through a small angle so that the direction of the particle is always close to the z-direction, the total change in v_x is

$$\Delta v_x = (\delta v_\perp)_1 \cos \theta_1 + (\delta v_\perp)_2 \cos \theta_2 + \cdots + (\delta v_\perp)_N \cos \theta_N.$$

The $\delta v_\perp$s will be distributed depending on the distribution of impact

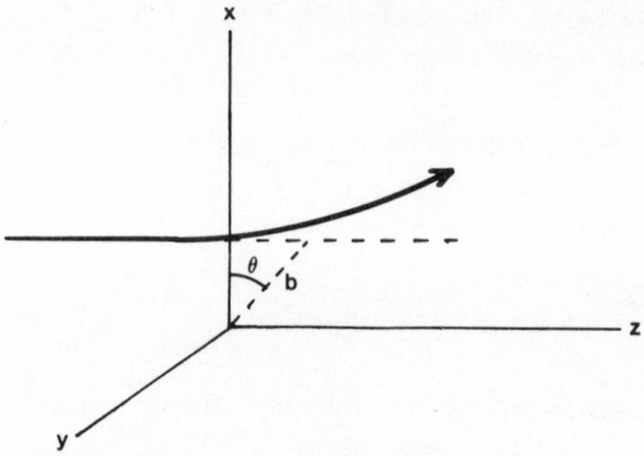

Figure 3.9 Geometry of collision process.

parameters, while the θs will be distributed evenly between 0 and 2π. Taking the average of Δv_x leads to zero, since the average of $\cos\theta$ vanishes. However the average of $(\Delta v_x)^2$ is non-zero, being

$$\langle \Delta v_x^2 \rangle = \tfrac{1}{2} N \langle \delta v_\perp^2 \rangle.$$

If these collisions take place in a time Δt, then

$$\frac{\langle \Delta v_x^2 \rangle}{\Delta t} = \tfrac{1}{2} R \langle \delta v_\perp^2 \rangle,$$

R being the average rate of collisions. Now, for a single collision

$$\delta v_\perp = \frac{\mu_{ss'}}{m_s} u \sin\psi \approx \frac{\mu_{ss'}}{m_s} u\psi,$$

and so taking the average over impact parameters and over particle velocity distributions as before we obtain

$$\frac{\langle \Delta v_x^2 \rangle}{\Delta t} = \frac{q_s^2}{4\pi\varepsilon_0^2 m_s^2} \sum_{s'} q_{s'}^2 \log\Lambda \int \frac{f_{s'}(\boldsymbol{v}')}{u} \mathrm{d}^3 v'.$$

It is clear that $\langle \Delta v_y^2 \rangle$ is the same thing, while $\langle \Delta v_x \Delta v_y \rangle$ is zero, since it involves the average of $\cos\theta \sin\theta$. Similarly $\langle \Delta v_x \Delta v_z \rangle$ and $\langle \Delta v_y \Delta v_z \rangle$ are zero, while $\langle \Delta v_z \Delta v_z \rangle$ depends on ψ^4 and so is very small for small-angle collisions. We shall approximate this last quantity by zero. Thus there are only two diagonal components of $\langle \Delta v_i \Delta v_j \rangle$ which are non-zero in this co-ordinate system. The corresponding property in a general coordinate system is reproduced by the formula

$$\begin{aligned} \frac{\langle \Delta v_i \Delta v_j \rangle}{\Delta t} &= \frac{q_s^2}{4\pi\varepsilon_0^2 m_s^2} \sum_{s'} q_{s'}^2 \log\Lambda \int \left(\delta_{ij} - \frac{u_i u_j}{u^2} \right) \frac{f_{s'}(\boldsymbol{v}')}{u} \mathrm{d}^3 v'. \\ &= 2D_{ij}. \end{aligned} \tag{3.65}$$

From (3.64) and (3.65), the coefficient A_i to appear in the Fokker–Plank equation is found to be

$$A_i = -\frac{q_s^2}{8\pi\varepsilon_0^2 m_s}\sum_{s'} q_{s'}^2 \log \Lambda \int\left[\frac{2u_i}{\mu_{ss'}u^3} + \frac{1}{m_s}\frac{\partial}{\partial v_j}\left(\frac{\delta_{ij}}{u} - \frac{u_i u_j}{u^3}\right)\right] f_{s'}(\mathbf{v}')\mathrm{d}^3 v'.$$

We can simplify this by noting that

$$\frac{\delta_{ij}}{u} - \frac{u_i u_j}{u^3} = \frac{\partial^2 u}{\partial u_i \partial u_j},$$

$$\frac{\partial}{\partial v_j}\left(\frac{\delta_{ij}}{u} - \frac{u_{ij}}{u^3}\right) = \frac{\partial}{\partial u_j}\left(\frac{\delta_{ij}}{u} - \frac{u_i u_j}{u^3}\right) = -\frac{2u_i}{u^3}, \tag{3.66}$$

and

$$\frac{1}{\mu_{ss'}} = \frac{1}{m_s} + \frac{1}{m_{s'}},$$

which can be used to give

$$A_i = -\frac{q_s^2}{4\pi\varepsilon_0^2 m_s}\sum_{s'}\frac{q_{s'}^2 \log \Lambda}{m_{s'}}\int \frac{u_i}{u^3} f_{s'}(\mathbf{v}')\,\mathrm{d}^3 v'. \tag{3.67}$$

Since $\mathbf{u} = \mathbf{v} - \mathbf{v}'$

$$\frac{\partial}{\partial v_j'}\left(\frac{\delta_{ij}}{u} - \frac{u_i u_j}{u^3}\right) = -\frac{\partial}{\partial v_j}\left(\frac{\delta_{ij}}{u} - \frac{u_i u_j}{u^3}\right) = \frac{2u_i}{u^3},$$

so that

$$A_i = -\frac{q_s^2}{8\pi\varepsilon_0^2 m_s}\sum_{s'}\frac{q_{s'}^2 \log \Lambda}{m_s'}\int \frac{\partial}{\partial v_j'}\left(\frac{\delta_{ij}}{u} - \frac{u_i u_j}{u^3}\right) f_{s'}(\mathbf{v}')\,\mathrm{d}^3 v'.$$

An integration by parts then gives

$$A_i = \frac{q_s^2}{8\pi\varepsilon_0^2 m_s}\sum_{s'}\frac{q_{s'}^2 \log \Lambda}{m_{s'}}\int \left(\frac{\delta_{ij}}{u} - \frac{u_{ij}}{u^3}\right)\frac{\partial f_{s'}(\mathbf{v}')}{\partial v_j'}\,\mathrm{d}^3 v'.$$

Combining this with the expression already found for D_{ij} gives the kinetic equation in the form

$$\frac{\partial f_s(\mathbf{v})}{\partial t} = -\sum_{s'}\frac{q_s^2 q_{s'}^2 \log \Lambda}{8\pi\varepsilon_0^2 m_s}\frac{\partial}{\partial v_i}\int\left(\frac{\delta_{ij}}{u} - \frac{u_i u_j}{u^3}\right)\left\{\frac{f_s(\mathbf{v})}{m_s'}\frac{\partial f_{s'}(\mathbf{v}')}{\partial v_j'} - \frac{f_{s'}(\mathbf{v}')}{m_s}\frac{\partial f_s(\mathbf{v})}{\partial v_j}\right\}\mathrm{d}^3 v. \tag{3.68}$$

The right-hand side of this version of the kinetic equation is usually known as the Landau collision integral.

An alternative form which is commonly used in applications can be obtained by using (3.67) for A and (3.66) in the expression (3.65) for D_{ij}. The result is

$$\frac{\partial f_s(\mathbf{v})}{\partial t} = -\sum_{s'}\frac{q_s^2 q_{s'}^2 \log \Lambda}{4\pi\varepsilon_0^2 m_s^2}\frac{\partial}{\partial v_i}\left[f_s(\mathbf{v})\frac{\partial}{\partial v_i}h(\mathbf{v}) - \frac{1}{2}\frac{\partial}{\partial v_j}\left(f_s(\mathbf{v})\frac{\partial^2 g}{\partial v_i \partial v_j}\right)\right], \tag{3.69}$$

where g and h are the so-called Rosenbluth potentials,

$$g(\boldsymbol{v}) = \int f_{s'}(\boldsymbol{v}')u\,\mathrm{d}^3v'$$

$$h(\boldsymbol{v}) = \left(1 + \frac{m_s}{m_{s'}}\right)\int f_{s'}(\boldsymbol{v}')u^{-1}\,\mathrm{d}^3v'. \tag{3.70}$$

Equations (3.68) and (3.69) are both of the general form of a Fokker–Plank equation and are the most commonly used of all the various ways of writing the collision term. They differ from the Fokker–Plank equations as applied to Brownian motion in that the friction and diffusion coefficients themselves depend on the distribution functions, making the equation nonlinear. In many applications it is linearized by using Maxwellian distributions in the Rosenbluth potentials, an approximation which is reasonable if the distributions are not too far from thermal equilibrium.

The Fokker–Plank collision term derived above can be shown, probably most conveniently from the version given in (3.68), to conserve particles, momentum and energy. Thus

$$\frac{\partial}{\partial t}\int f_s\,\mathrm{d}^3v = \frac{\partial}{\partial t}\sum_s m_s \int \boldsymbol{v} f_s\,\mathrm{d}^3v$$

$$= \frac{\partial}{\partial t}\sum_s \tfrac{1}{2}m_s \int v^2 f_s\,\mathrm{d}^3v = 0. \tag{3.71}$$

These conservation laws are clearly a necessary property of any physically correct collision term. Another important property is that a state in which each species has a Maxwellian distribution with common average velocity and temperature is an equilibrium solution.

3.8 Relation between kinetic and fluid theories

A kinetic equation such as the Fokker–Plank equation provides a description of the plasma on a microscopic level where the distribution of particles is known, while the fluid equations deal only in macroscopic quantities like density, mean velocity, etc., which are moments of the velocity distribution functions. In this section we shall look briefly at the problem of deriving fluid from kinetic equations.

Since fluid equations deal with an inhomogeneous plasma in general, we must first note that the generalization of (3.68) and (3.69) to an inhomogeneous system containing electric or magnetic fields is

$$\frac{\partial f_s}{\partial t} + \boldsymbol{v}\cdot\frac{\partial f_s}{\partial \boldsymbol{r}} + \frac{q_s}{m_s}(\boldsymbol{E} + \boldsymbol{v}\times\boldsymbol{B})\cdot\frac{\partial f_s}{\partial \boldsymbol{v}} = \text{collision term}, \tag{3.72}$$

the collision term on the right-hand side being as before. On the left-hand side $\partial/\partial \boldsymbol{r}$ is the gradient operator in ordinary space and $\partial/\partial \boldsymbol{v}$ the corresponding operator in velocity space, i.e.

$$\frac{\partial f_s}{\partial \boldsymbol{v}} = \left(\frac{\partial f_s}{\partial v_x}, \frac{\partial f_s}{\partial v_y}, \frac{\partial f_s}{\partial v_z}\right),$$

while the whole thing is simply the total derivative following a particle orbit in the six-dimensional phase space with coordinates $(\boldsymbol{r}, \boldsymbol{v})$. It can be regarded as a generalization of the familiar convective derivative of fluid mechanics.

Since the fluid quantities are moments of the distribution function we might try to obtain fluid equations by taking moments of the kinetic equation. So, for instance, if we integrate (3.72) over velocity we get

$$\frac{\partial n_s}{\partial t} + \boldsymbol{\nabla}\cdot(n_s \boldsymbol{u}_s) = 0, \tag{3.73}$$

where $\boldsymbol{u}_s$ is the mean velocity of the species s, defined by

$$n_s \boldsymbol{u}_s = \int \boldsymbol{v} f_s \, \mathrm{d}^3 v.$$

To find the rate of change of $\boldsymbol{u}_s$ we may multiply (3.72) by $\boldsymbol{v}$ and integrate. However, the second term on the left-hand side then introduces the divergence of the tensor $\int v_i v_j f \, \mathrm{d}^3 v$ which is related to the pressure tensor. In the same way the rate of change of successively higher-order moments of f each involves the gradient of a still higher-order moment, so that the set of moment equations is not closed. The equations may be closed by making some *ad hoc* assumptions, as for example assuming an isotropic pressure and zero heat flow, which will give us the set of equations used in the earlier part of this chapter.

A more systematic procedure may be followed using the method of Chapman and Enskog, which was first developed in relation to the kinetic theory of neutral gases. This is a systematic perturbation method, based on the assumption that the length and time scales for variation of the plasma are long compared to the collision frequency or mean free path. (In a plasma where a single collision results on average in only a very small change in velocity, the latter are usually taken to be the average time or distance required for the velocity of a particle to be deflected through a right angle.) In this case the left-hand side of the kinetic equation is small. If the distribution function of each species is expanded in a series

$$f_s = f_s^{(0)} + f_s^{(1)} + f_s^{(2)} + \cdots,$$

then to lowest order the left-hand side of the kinetic equation is ignored and the condition determining $f_s^{(0)}$ is just that the collision term vanishes. This imposes the condition that each $f_s^{(0)}$ is Maxwellian, of the form

$$f_s^{(0)} = \frac{n_s}{(2\pi\kappa T/m_s)^{3/2}} \, \mathrm{e}^{-m_s(\boldsymbol{v}-\boldsymbol{u})^2/2\kappa T},$$

where the quantities n_s, $\boldsymbol{u}$ and T may be slowly varying functions of $\boldsymbol{r}$ and t.

In the next order in the expansion f_s is approximated by $f_s^{(0)}$ on the left-hand side of the kinetic equation, which then involves gradients of the densities, mean velocity and temperature. On the right-hand side f_s is approximated by $f_s^{(0)}+f_s^{(1)}$ and terms linear in $f_s^{(1)}$ are kept. In principle this allows calculation of $f_s^{(1)}$ in terms of n_s, $\boldsymbol{u}$ and T. Moment equations giving the rate of change of n_s, $\boldsymbol{u}$ and T can be constructed, and any other moments which appear in them be calculated from the distribution functions correct to first order, so that a closed set of equations in n_s, $\boldsymbol{u}$, T and their derivatives is constructed—the fluid equations which we wish. The Chapman–Enskog procedure thus allows a systematic derivation of a closed set of fluid equations from the kinetic equations, though the above should only be regarded as giving the basic idea of a procedure which in detail is technically complicated.

3.9 Transport coefficients

We have seen in the last section how the Chapman–Enskog procedure allows calculation of moments of the distribution function like the pressure tensor and the heat flow in terms of gradients of density, velocity and temperature. Now, in the standard theory of fluids the force on the fluid, or contribution to the pressure tensor, due to a velocity gradient is ascribed to viscosity, and the coefficient of viscosity relates the gradient to the pressure. The Chapman–Enskog procedure thus allows a calculation of the coefficient of viscosity from the collision integral. In this way the kinetic theory outlined in the last few sections allows a calculation of the coefficients of viscosity and other transport properties such as diffusion and thermal conductivity in terms of the basic properties of the inter-particle interactions.

Because of the complicated form of the Fokker–Plank collision term numerical techniques are required to solve it for $f_s^{(1)}$ and obtain the moments which determine the transport coefficients. However we may illustrate the general idea and obtain results, which, if not numerically exact, have the correct dependence on the plasma parameters, by using a simplified theory. In this we replace the collision term with

$$-\nu(f-f_0),$$

where ν is a collision frequency and f_0 is the Maxwellian distribution with the same density, mean velocity and mean energy as are given by taking moments of f.

If we consider first the electrical conductivity, we are looking for the electron response to a steady electric field, so with our simplified collision term we have to solve the equation

$$-\frac{e}{m}\boldsymbol{E}\cdot\frac{\partial f}{\partial \boldsymbol{v}} = -\nu(f-f_0).$$

With the unperturbed plasma assumed to be at rest the zero-order approximation involves equating the right-hand side to zero, so that

$$f=f_0=\frac{n_e}{(2\pi\kappa T/m_e)^{3/2}}\,\mathrm{e}^{-1/2m_e v^2/\kappa T}$$

Putting $f=f_0+f_1$ and putting $f=f_0$ on the left-hand side gives

$$-\frac{e}{m}\boldsymbol{E}\cdot\frac{\partial f_0}{\partial \boldsymbol{v}}=-\nu f_1,$$

and so

$$\boldsymbol{J}=\int \boldsymbol{v} f_1\,\mathrm{d}^3 v=-\frac{e^2}{m_e\nu}\int \boldsymbol{v}\left(\boldsymbol{E}\cdot\frac{\partial f_0}{\partial \boldsymbol{v}}\right)\mathrm{d}^3 v$$

$$=\frac{e^2}{m_e\nu}\int \boldsymbol{E} f_0\,\mathrm{d}^3 v=\frac{n_e e^2}{m_e\nu}\boldsymbol{E}.$$

The electrical conductivity is thus

$$\sigma=\frac{n_e e^2}{m_e\nu}. \tag{3.74}$$

Since the electrical resistance of a plasma arises from electron–ion collisions we may use for ν a characteristic collision frequency for electron–ion collisions. Estimating the electron slowing-down time owing to collisions with ions from, for example, equation (3.64) suggests that we take

$$\nu\approx\frac{Z^2 e^4 n_i \log\Lambda}{4\pi\varepsilon_0^2 m_e^{1/2}(\kappa T)^{3/2}}$$

giving

$$\sigma=\frac{4\pi\varepsilon_0^2(\kappa T)^{3/2}}{Ze^2 m_e^{1/2}\log\Lambda}. \tag{3.75}$$

This agrees with our earlier estimate of resistivity (3.60), at least in so far as dependence on the plasma parameters is concerned. To obtain the correct numerical factor requires solution of the Fokker–Plank equation by numerical methods.

As a second example of the derivation of a transport coefficient we consider the electron diffusion coefficient, which is the ratio of particle flux to density gradient. With the temperature taken to be constant we have

$$f_0=n_e(\boldsymbol{r})\left(\frac{m_e}{2\pi\kappa T}\right)^{3/2}\mathrm{e}^{-m_e v^2/2\kappa T},$$

and the first-order correction is determined by

$$\boldsymbol{v}\cdot\frac{\partial f_0}{\partial \boldsymbol{r}}=-\nu f_1.$$

The particle flux is

$$\begin{aligned}\Gamma &= \int \boldsymbol{v} f_1 \, \mathrm{d}^3 v \\ &= -\frac{1}{n_e \nu} \int f_0 \left(\boldsymbol{v} \cdot \frac{\partial n}{\partial \boldsymbol{r}} \right) \boldsymbol{v} \, \mathrm{d}^3 v \\ &= -\frac{\kappa T}{m_e \nu} \boldsymbol{\nabla} n,\end{aligned}$$

giving us an estimate of the diffusion coefficient

$$D = \frac{\kappa T}{m_e \nu}. \tag{3.76}$$

The inverse dependence on the mass means that the ion diffusion coefficient is much smaller.

In a plasma with a density gradient, both ions and electrons tend to diffuse down the density gradient, but with the electrons moving faster. This, however, would tend to lead to a charge separation, and except on length scales of the order of a Debye length, any significant charge separation leads to a very large electric field.

As shown in Fig. 3.10 an electric field develops tending to hold back the electrons and speed up the ions. Unless the gradients are very steep the electric field adjusts itself to maintain 'quasineutrality', that is a state in which electron and ion charge densities are very nearly equal. This means that electrons and ions flow at the same rate, in a process known as ambipolar diffusion. In problems involving long scale-lengths the assumption of quasineutrality is often used to determine the electric field, in place of Poisson's equation.

The electron and ion fluxes are

$$\begin{aligned}\boldsymbol{\Gamma}_i &= -D_i \boldsymbol{\nabla} n_i + n_i \mu_i \boldsymbol{E} \\ \boldsymbol{\Gamma}_e &= -D_e \boldsymbol{\nabla} n_e + n_e \mu_e \boldsymbol{E},\end{aligned} \tag{3.77}$$

where μ_e and μ_i are the particle mobilities, defined as the average particle velocity produced per unit electric field. For electrons

$$\mu_e = \frac{\sigma}{n_e e} = -\frac{e}{m_e \nu}$$

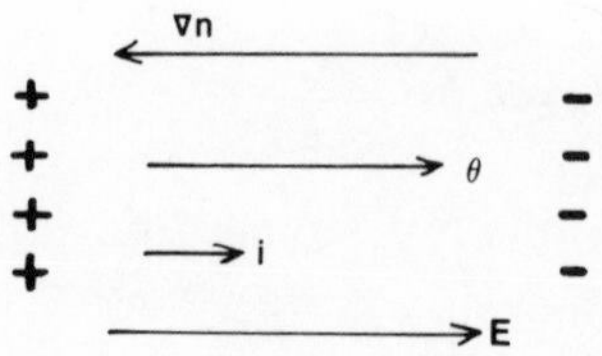

Figure 3.10 Ambipolar diffusion.

and similarly

$$\mu_i = \frac{Ze}{m_i v}.$$

Taking $Z = 1$ for simplicity, we must have $n_e = n_i$ and also $\Gamma_e = \Gamma_i$. Imposing these conditions we may eliminate E from the pair of equations and use the fact that $\mu_e/\mu_i = -D_e/D_i$ to obtain

$$\boldsymbol{\Gamma}_i = \boldsymbol{\Gamma}_e = -D_a \boldsymbol{\nabla} n_e = -D_a \boldsymbol{\nabla} n_i,$$

where D_a is the ambipolar diffusion coefficient given by

$$D_a = \frac{2D_e D_i}{D_e + D_i} \approx 2D_i.$$

The rate of diffusion of the plasma is thus controlled by that of the ions.

The transport coefficients considered above do not include the effect of a steady magnetic field in the plasma, so we shall now look at diffusion in the presence of a magnetic field. In view of the use of magnetic fields to confine plasma, the rate of diffusion across the field is obviously of the utmost importance. The zero order distribution function can again be taken to be Maxwellian, since $(\boldsymbol{v} \times \boldsymbol{B})\cdot\partial f_0/\partial \boldsymbol{v} = 0$ for this distribution. Then, taking the term in $(\boldsymbol{v} \times \boldsymbol{B})\cdot(\partial f_1/\partial \boldsymbol{v})$ into account in the equation for the first-order correction to the electron distribution function, we have

$$\begin{aligned}\frac{f_0}{n_e}\boldsymbol{v}\cdot\frac{\partial n}{\partial \boldsymbol{r}} &= -\nu f_1 + \frac{e}{m_e}(\boldsymbol{v} \times \boldsymbol{B})\cdot\frac{\partial f_1}{\partial \boldsymbol{v}} \\ &= -\nu f_1 + \Omega_e v_y \frac{\partial f_1}{\partial v_x} - \Omega_e v_x \frac{\partial f_1}{\partial v_y},\end{aligned}$$

where $\Omega_e = eB/m_e$ and the magnetic field has been taken to be along the z-axis.

Multiplying this by v_x, v_y, v_z in turn and integrating over velocity gives

$$\frac{\kappa T}{m_e}\frac{\partial n}{\partial x} = -\nu\Gamma_x - \Omega_e\Gamma_y$$

$$\frac{\kappa T}{m_e}\frac{\partial n}{\partial y} = -\nu\Gamma_y + \Omega_e\Gamma_x$$

$$\frac{\kappa T}{m_e}\frac{\partial n}{\partial z} = -\nu\Gamma_z,$$

which results in

$$\Gamma_i = -D_{ij}\frac{\partial n}{\partial r_j},$$

with the diffusion tensor D given by

$$D_{ij} = -\frac{\kappa T}{m_e}\begin{pmatrix} \dfrac{\nu}{\nu^2+\Omega_e^2} & -\dfrac{\Omega_e}{\nu^2+\Omega_e^2} & 0 \\ \dfrac{\Omega_e}{\nu^2+\Omega_e^2} & \dfrac{\nu}{\nu^2+\Omega_e^2} & 0 \\ 0 & 0 & \dfrac{1}{\nu} \end{pmatrix} \tag{3.78}$$

From this it can be seen that the electron diffusion along a density gradient perpendicular to a magnetic field is reduced by a factor $\nu^2/(\nu^2+\Omega_e^2)$ compared to that in the absence of a field. In most confinement experiments $\nu \ll |\Omega_e|$ and so the diffusion across the field goes as $1/B^2$ and is proportional to ν. The physical reason for the direct proportionality of the diffusion and collision frequency is that in the absence of collisions the particle is tied to its orbit about a magnetic field line and does not move across the field at all. Collisions can scatter it to an orbit about a different field line and so lead to its gradual diffusion across the field. The $1/\Omega_e^2$ dependence means that the corresponding term for ions is larger, reversing the ordering in the absence of the field. This is simply a consequence of the larger ion Larmor radius.

The above diffusion tensor applies to straight magnetic field lines and in a confined plasma does not necessarily apply. In a uniform field a collision (interpreted as a 90° deflection) produces a jump in position of a particle of the order of a Larmor radius. However in section 2.5 we have looked at the nature of particle orbits in a tokamak. Untrapped particles, which go right round the tokamak, deviate from the magnetic surface as illustrated in Figure 2.11, and a collision may move a particle by a distance of the order of this deviation. Also, if trapped particles are subject to collision they may move across the field by a distance of the order of the width of their banana orbit. In this way the diffusion in a tokamak (and in other toroidal devices) is greater than that in a uniform field. A sophisticated theory, known as neoclassical theory, has been developed to predict the effects of toroidal geometry on collisional transport processes.

Unfortunately, although neoclassical diffusion is the subject of very elegant theoretical analysis, it has not proved capable of explaining experimental results. Energy transport across the field by electrons is found to be some two orders of magnitude greater than predicted. It is generally accepted that the reason is that the plasma is not in a stable thermal equilibrium state, but is subject to instabilities which produce an enhanced level of fluctuations. These lead in turn to an enhancement of the diffusion rate. Theoretical analysis of such effects is, however, an extremely difficult matter, and at present this topic, which is of central importance in the study of magnetic confinement, is the subject of a great deal of research.

We have discussed the Fokker–Plank equations mainly as a step towards obtaining the transport coefficients to be inserted in fluid theories. There are, however, some problems which require direct solution of the Fokker–Plank equation. One such occurs in laser-produced plasmas where very steep temperature gradients are produced. These are so steep that the normal theory of thermal conduction breaks down and heat flow has to be computed from numerical solutions of the Fokker–Plank equation.

Problems

3.1 Derive equation 3.19.

3.2 If, in equation (3.14), the pressure gradient can be neglected, show that

$$\nabla \times \boldsymbol{B} = \alpha \boldsymbol{B},$$

where α may be an arbitrary function of position. Fields satisfying this condition are called 'force-free' and are important in some astrophysical contexts, for example the Sun's corona.

If α is constant show that $\boldsymbol{B}$ obeys the Helmholtz equation

$$(\nabla^2 + \alpha^2)\boldsymbol{B} = 0.$$

3.3 Flute-like modes on a cylindrical plasma column, with the perturbations aligned along the external electric field, have $kB_e + (m/a)B_\theta(a) = 0$. Show that, in the limit of long wavelength, the criterion for instability of such modes is $B_e^2 > mB_i^2$.

3.4 Derive (3.66) and the equation which follows it.

3.5 Show that the Kruskal–Shafranov condition applied to a tokamak implies that for stability the safety factor must be greater than one.

3.6 Obtain the Fokker–Plank equation in the form of equation (3.69).

3.7 Using the model described in section 3.9 show that the thermal conductivity due to electrons in a plasma with constant pressure is

$$K = \frac{5n_e\kappa^2 T}{2m_e\nu}.$$

3.8 A slab of plasma of uniform pressure lies between the planes $x = 0$ and $x = a$, the temperature being T_0 at the former and T_1 at the latter. Show that the temperature profile between the planes is given by

$$\frac{T_0^{7/2} - T^{7/2}}{T_0^{7/2} - T_1^{7/2}} = \frac{x}{a}.$$

4 Waves

4.1 Introduction

In the last chapter we discussed the small-amplitude waves which are predicted by the equations of magnetohydrodynamics. The behaviour found there is, however, applicable only to low frequencies since the waves involve bulk motions of the whole plasma, with electrons and ions moving together and no net charge density appearing. At high frequencies we might expect oscillations in which the electrons oscillate, while the ions, which are much heavier, remain almost stationary. In this case charge separation may occur giving rise to electric fields, a phenomenon which the magnetohydrodynamic approximation, by its very nature, is incapable of describing. In section 4.2 we introduce a description of the plasma in which electrons and ions are treated as two distinct fluids, allowing just this sort of effect to be treated.

After a discussion of some of the properties of waves in a spatially uniform plasma we consider something of what happens when waves propagate in an inhomogeneous plasma. This is a subject of immense practical importance, since plasmas which occur either in the laboratory or in space generally have gradients in magnetic field, density or other properties. In section 4.6 we consider some aspects of the use of radio-frequency waves to heat tokamak plasmas, an important topic which provides an illustration of the preceding theory. The final section of the chapter is largely devoted to a study of resonant absorption, which illustrates some of the more complex properties of wave propagation in an inhomogeneous system and plays a significant role in the interaction of laser light with a plasma.

4.2 The two-fluid equations

We shall begin by allowing for an arbitrary number of species, recognizing that a plasma may contain more than one type of ion, but shall consider the detailed properties of only a plasma with a single ion species. The essential idea is to describe each species as a separate fluid, obeying equations of a type similar to these we have seen in Chapter 3. If we neglect the effect of plasma pressure, so that only the electromagnetic forces are taken into account, we obtain what is called the 'cold-plasma approximation'. Some of the limitations of this and the effect of finite temperature will be discussed later in this chapter and in the next chapter.

The basic equations are the equations of conservation of particle number

$$\frac{\partial n_s}{\partial t} + \nabla \cdot (n_s \boldsymbol{v}_s) = 0, \tag{4.1}$$

where n_s is the number of particles per unit volume and $\boldsymbol{v}_s$ the fluid velocity of the species s ($s = e$ for electrons and i for ions when we wish to distinguish them), and the momentum equations

$$\frac{\partial \boldsymbol{v}_s}{\partial t} + (\boldsymbol{v}_s \cdot \nabla)\boldsymbol{v}_s = \frac{q_s}{m_s}(\boldsymbol{E} + \boldsymbol{v}_s \times \boldsymbol{B}), \tag{4.2}$$

with q and m the particle charge and mass. There is one such set of equations for each species. In addition we must calculate the fields via Maxwell's equations, with the current density given by

$$\boldsymbol{J} = \sum_s q_s n_s \boldsymbol{v}_s. \tag{4.3}$$

The fields are, of course, the same in each set of fluid equations and, being dependent on the combined motion of all the particles, provide the link between the separate equations for the different species.

As was the case when we studied magnetohydrodynamic waves we shall consider small perturbations about an initial state of uniform density. If there is to be overall charge neutrality the initial densities n_{0s} must satisfy the condition $\Sigma n_{0s} q_s = 0$. We shall assume that each species is at rest in the unperturbed state, in which we allow a uniform magnetic field $\boldsymbol{B}_0$ but no electric field. Linearizing the equations, in the same manner as discussed in Chapter 3, with the perturbations distinguished by a subscript 1, we obtain from (4.1) and (4.2)

$$\frac{\partial n_{1s}}{\partial t} + n_{0s} \nabla \cdot \boldsymbol{v}_{1s} = 0 \tag{4.4}$$

and

$$\frac{\partial \boldsymbol{v}_{1s}}{\partial t} = \frac{q_s}{m_s}(\boldsymbol{E}_1 + \boldsymbol{v}_{1s} \times \boldsymbol{B}_0), \tag{4.5}$$

while Maxwell's equations give us

$$\nabla \times \boldsymbol{E}_1 = -\frac{\partial \boldsymbol{B}_1}{\partial t} \tag{4.6}$$

$$\nabla \times \boldsymbol{B}_1 = \sum_s \mu_0 n_{0s} q_s \boldsymbol{v}_{1s} + \frac{1}{c^2}\frac{\partial \boldsymbol{E}_1}{\partial t} \tag{4.7}$$

In Maxwell's equations we use the free-space permittivity and permeability, since the effects of the plasma are included explicitly via the direct calculation of the current, rather than implicitly through a dielectric constant or relative permeability different from unity. The reason for the neglect of the remaining two Maxwell's equations is explained in Problem 4.1.

Having linearized the equations, the next step in the investigation of waves is to look for a solution in which all variables go as $\exp(i\boldsymbol{k}\cdot\boldsymbol{r} - i\omega t)$, representing a plane wave of angular frequency ω and wavenumber $\boldsymbol{k}$. An arbitrary perturbation is capable of representation, through Fourier analysis, as a superposition of such solutions. With this assumption (4.5) becomes

$$-i\omega\boldsymbol{v}_{1s} = \frac{q_s}{m_s}(\boldsymbol{E}_1 + \boldsymbol{v}_{1s} \times \boldsymbol{B}_0), \tag{4.8}$$

where we now use the same symbol $\boldsymbol{v}_{1s}$ to mean the amplitude of the plane wave solution. Taking the vector and scalar products of (4.8) with $\boldsymbol{B}_0$ and eliminating $\boldsymbol{v}_{1s} \times \boldsymbol{B}_0$ and $\boldsymbol{v}_{1s}\cdot\boldsymbol{B}_0$ from the resulting set of equations gives

$$\boldsymbol{v}_{1s} = \frac{iq_s}{m_s\omega(1 - \Omega_s^2/\omega^2)}\left[\boldsymbol{E}_1 + \frac{i\Omega_s}{\omega}(\boldsymbol{E}_1 \times \hat{\boldsymbol{b}}) - \frac{\Omega_s^2}{\omega^2}(\boldsymbol{E}_1\cdot\hat{\boldsymbol{b}})\hat{\boldsymbol{b}}\right], \tag{4.9}$$

where $\hat{\boldsymbol{b}}$ is a unit vector along the direction of $\boldsymbol{B}_0$ and $\Omega_s = q_s B_0/m_s$ is the cyclotron frequency of the species s, with negative sign in the case of electrons.

Eliminating $\boldsymbol{B}_1$ from (4.6) and (4.7) in which the gradient operator is now replaced with $i\boldsymbol{k}$, gives

$$\boldsymbol{k} \times (\boldsymbol{k} \times \boldsymbol{E}_1) = -i\omega\mu_0 n_0 \sum q_s \boldsymbol{v}_{1s} - \frac{\omega^2}{c^2}\boldsymbol{E}_1 \tag{4.10}$$

which with the aid of (4.9) can be separated into components consisting of a set of three linear equations in E_{1x}, E_{1y} and E_{1z}. The condition that there exists a non-trivial solution is that the determinant of the coefficients vanishes, and this condition gives an equation relating ω and $\boldsymbol{k}$, known as the dispersion relation. For waves of a given wavenumber this determines the possible wave frequencies, and gives us details of their phase and group velocities, $(\omega/k)\hat{\boldsymbol{k}}$ and $\partial\omega/\partial\boldsymbol{k}$ respectively.

Instead of $\boldsymbol{k}$ it is convenient to introduce the dimensionless vector $\boldsymbol{n} = \boldsymbol{k}c/\omega$. Also, we adopt the coordinate system which is standard in this subject, with the steady magnetic field in the z-direction and the x-axis along the component of $\boldsymbol{k}$ (or $\boldsymbol{n}$) perpendicular to the field. If θ is the angle between $\boldsymbol{n}$ and $\boldsymbol{B}$, (4.10) takes the form

$$\begin{pmatrix} \varepsilon_{11} - n^2\cos^2\theta & \varepsilon_{12} & n^2\cos\theta\sin\theta \\ \varepsilon_{21} & \varepsilon_{22} - n^2 & 0 \\ n^2\cos\theta\sin\theta & 0 & \varepsilon_{33} - n^2\sin^2\theta \end{pmatrix}\begin{pmatrix} E_{1x} \\ E_{1y} \\ E_{1z} \end{pmatrix} = 0 \tag{4.11}$$

in which the terms involving n come from the left-hand side of (4.10), while the ε_{ij}, the elements of the plasma dielectric tensor, come from the right-hand side and are given by

$$\varepsilon_{11} = \varepsilon_{22} = 1 - \sum_s \frac{\omega_{ps}^2}{\omega^2 - \Omega_s^2}$$

$$\varepsilon_{12} = -\varepsilon_{21} = -i\sum_s \frac{\omega_{ps}^2 \Omega_s}{\omega(\omega^2 - \Omega_s^2)} \qquad (4.12)$$

$$\varepsilon_{33} = 1 - \sum_s \frac{\omega_{ps}^2}{\omega^2},$$

where $\omega_{ps} = (n_{0s} q_s^2/\varepsilon_0 m_s)^{1/2}$ is known as the plasma frequency of the species s.

4.3 Waves in a cold plasma

The dispersion relation obtained by setting the determinant of the coefficients of (4.11) to zero contains, in principle, all the information about the different wave modes which may propagate in the plasma, at least within the limitations of our cold-plasma model, but not in a very transparent form. To get some idea of the variety of waves which are possible we shall analyse various special cases, beginning with the simplest of all, that when the magnetic field vanishes. In this case

$$\varepsilon_{11} = \varepsilon_{22} = \varepsilon_{33} = 1 - \frac{\omega_p^2}{\omega^2}, \quad \text{where} \quad \omega_p^2 = \sum_s \omega_{ps}^2, \quad \text{and} \quad \varepsilon_{12} = \varepsilon_{21} = 0.$$

Since the electron mass is very much smaller than the ion mass, the plasma frequency ω_p is very close to the electron plasma frequency. With no magnetic field there is no preferred direction, so with no loss of generality we may take $\boldsymbol{k}$ to be in the z-direction, obtaining

$$\begin{pmatrix} 1 - \dfrac{\omega_p^2}{\omega^2} - n^2 & 0 & 0 \\ 0 & 1 - \dfrac{\omega_p^2}{\omega^2} - n^2 & 0 \\ 0 & 0 & 1 - \dfrac{\omega_p^2}{\omega^2} \end{pmatrix} \begin{pmatrix} E_{1x} \\ E_{1y} \\ E_{1z} \end{pmatrix} = 0. \qquad (4.13)$$

From this it is clear that there are two types of wave, one with $E_{1z} \neq 0$ which has the dispersion relation

$$1 - \frac{\omega_p^2}{\omega^2} = 0, \qquad (4.14)$$

and the other with E_{1x} or $E_{1y} \neq 0$ which has the dispersion relation

$$1 - \frac{\omega_p^2}{\omega^2} - \frac{k^2 c^2}{\omega^2} = 0. \qquad (4.15)$$

The first of these has $\boldsymbol{k}$ parallel to $\boldsymbol{E}_1$, and so is a longitudinal wave, known as the plasma wave, with a frequency of $\omega = \omega_p$. If $\boldsymbol{k}$ and $\boldsymbol{E}_1$ are parallel, then we can see from (4.6) that $\boldsymbol{B}_1$ vanishes, implying that the wave field is purely electrostatic. The physical mechanism producing the wave is quite simply that if an excess of electrons is created at some point the resulting electric field is such as to push them apart, producing a restoring force which can give rise to the oscillation. Since ω is independent of $\boldsymbol{k}$, the group velocity is zero, so this is not a propagating wave but has the property that an oscillation set up in one region of the plasma remains localized in this region. Perhaps we might point out that this is a property which does not obtain in a real plasma with a finite temperature, as will be seen in the following chapter. The motion is predominantly of the electrons, which is why $\omega_p \approx \omega_e$.

The other wave is a transverse mode with $\boldsymbol{k}$ perpendicular to $\boldsymbol{E}_1$, and with two mutually perpendicular polarizations corresponding to $E_{1x} \neq 0$ or $E_{1y} \neq 0$, just like a vacuum electromagnetic wave. Its dispersion relation (4.15) can be rearranged to read $\omega^2 = \omega_p^2 + k^2c^2$, showing that this is just the usual electromagnetic wave, which in a vacuum has $\omega^2 = k^2c^2$, modified by the presence of the plasma. An important property, which follows immediately, is that for propagation of the wave we must have $\omega \geqslant \omega_p$. Since ω_p is proportional to the square root of the plasma density this means that light of a given frequency will only propagate in plasma below a certain critical density. This is important in the interaction of laser light with dense targets, ensuring that once the outer layer of the target is ionized the light must be absorbed in or reflected from the outer layer of the plasma which is below the critical density.

Now, we look at the less straightforward case where the magnetic field is non-zero, and the dispersion relation is

$$\begin{vmatrix} \varepsilon_{11} - n^2\cos^2\theta & \varepsilon_{12} & n^2\cos\theta\sin\theta \\ \varepsilon_{21} & \varepsilon_{22} - n^2 & 0 \\ n^2\cos\theta\sin\theta & 0 & \varepsilon_{33} - n^2\sin^2\theta \end{vmatrix} = 0 \tag{4.16}$$

If we regard ω as fixed and solve this for n and hence k, the resulting equation is a quadratic in n^2, with two roots, corresponding to two propagating wave modes if both are positive. The two roots of opposite sign for n corresponding to a root for n^2 simply correspond to waves of the same type propagating in opposite directions. Although it is possible, of course, to write down the solutions for n^2, the coefficients of the quadratic are so complicated that the properties are not clear. The alternative approach of finding ω for given k is much more difficult, and in any case it is generally the wave frequency which is fixed by the properties of the source. To give some idea of the wave modes which occur we shall examine some special cases.

First, let us look at frequencies well below the ion cyclotron or plasma frequencies, which are, in turn, well below the corresponding electron frequencies. From now on we shall specialize to a plasma with one ion species, singly charged so that $q_i = -q_e = e$. In this low-frequency regime we have

$$\varepsilon_{11} \approx 1 + \frac{\omega_{pe}^2}{\Omega_e^2} + \frac{\omega_{pi}^2}{\Omega_i^2} \approx 1 + \frac{\omega_{pi}^2}{\Omega_i^2}$$

$$\varepsilon_{12} = -\varepsilon_{21} \approx \frac{i}{\omega}\left(\frac{\omega_{pe}^2}{\Omega_e} + \frac{\omega_{pi}^2}{\Omega_i}\right) = 0$$

$$\varepsilon_{33} \approx 1 - \frac{\omega_{pe}^2}{\omega^2} \approx -\frac{\omega_{pe}^2}{\omega^2},$$

so that the dispersion relation is approximately

$$\begin{vmatrix} 1 + \dfrac{\omega_{pi}^2}{\Omega_i^2} - n^2 \cos^2\theta & 0 & n^2 \cos\theta \sin\theta \\ 0 & 1 + \dfrac{\omega_{pi}^2}{\Omega_i^2} - n^2 & 0 \\ n^2 \cos\theta \sin\theta & 0 & -\dfrac{\omega_{pe}^2}{\omega^2} - n^2 \sin^2\theta \end{vmatrix} = 0 \qquad (4.17)$$

In following the above remember that $\omega_{pe}^2 = (m_i/m_e)\omega_{pi}^2$ and $\Omega_e = -(m_i/m_e)\Omega_i$. If we now make the further observation that $\omega_{pe}^2/\omega^2 \gg \omega_{pi}^2/\Omega_i^2$ in the frequency range of interest, then we can see that the element in the bottom right-hand corner of the determinant is far larger than any of the others, so that a good approximation to the roots of the equation is obtained by equating to zero the term multiplying this large factor. Thus we obtain two roots

$$n^2 \cos^2\theta = 1 + \frac{\omega_{pi}^2}{\Omega_i^2}.$$

$$n^2 = 1 + \frac{\omega_{pi}^2}{\Omega_i^2}. \qquad (4.18)$$

From the definitions of the plasma and cyclotron frequencies it is a simple matter to show that

$$\frac{\omega_{pi}^2}{\Omega_i^2} = \frac{c^2}{(B_0^2/\mu_0\rho)}$$

where $\rho \approx n_0 m_i$ is the plasma density. This is just c^2/V_A^2 where V_A is the Alfvén speed introduced in the previous chapter. The equations (4.18) may now be rearranged to give the dispersion relations of the two low-frequency waves in the form

$$\omega = \frac{kV_A \cos\theta}{(1 + V_A^2/c^2)^{1/2}}$$

and

$$\omega = \frac{kV_A}{(1 + V_A^2/c^2)^{1/2}},$$

or if, as is generally the case, $V_A \ll c$, $\omega = kV_A \cos\theta$ and $\omega = kV_A$.

If we compare this with the results of the previous chapter on magnetohydrodynamic waves we see that the first solution is just the Alfvén wave, while the second is the zero-pressure limit of the fast magnetosonic wave. The slow magnetosonic wave disappears altogether as the pressure goes to zero. Names commonly used to denote these two low-frequency cold plasma waves are shear and compressional Alfvén waves or slow and fast Alfvén waves. The second terminology simply refers to the ordering of the phase velocities of the waves, while the first refers to the nature of the velocity fields associated with them.

For arbitrary frequencies there are two cases which are susceptible to simple analysis, namely parallel propagation, that is propagation along the field with $\theta = 0$, and perpendicular propagation, with $\theta = \pi/2$. If we consider the first of these then we see that the dispersion relation is

$$\begin{vmatrix} \varepsilon_{11} - n^2 & \varepsilon_{12} & 0 \\ \varepsilon_{21} & \varepsilon_{22} - n^2 & 0 \\ 0 & 0 & \varepsilon_{33} \end{vmatrix} = 0. \tag{4.19}$$

One possible solution is evidently to have $\varepsilon_{33} = 0$, i.e. $\omega^2 = \omega_p^2$, the corresponding solution for the electric field having $E_{1z} \neq 0$. This is just the plasma wave which we found previously in an unmagnetized plasma. If its wave vector is aligned with the field then particles oscillate parallel to $\boldsymbol{B}_0$ and do not feel any $\boldsymbol{v} \times \boldsymbol{B}$ force, with the result that the magnetic field has no effect.

The other possibility of satisfying (4.19) occurs when the 2×2 determinant involving the x and y components vanishes, corresponding to a transverse wave with non-zero E_{1x} and E_{1y} components. Using the formulae given earlier for the ε_{ij} this determinant can be shown to vanish when

$$n^2 - 1 + \frac{\omega_{pe}^2}{\omega^2 - \Omega_e^2} + \frac{\omega_{pi}^2}{\omega^2 - \Omega_i^2} = \pm\left[\frac{\omega_{pe}^2}{\omega^2 - \Omega_e^2}\frac{\Omega_e}{\omega} + \frac{\omega_{pi}^2}{\omega^2 - \Omega_i^2}\frac{\Omega_i}{\omega}\right]$$

which can be simplified to

$$1 - n^2 - \frac{\omega_{pe}^2}{\omega(\omega \pm \Omega_e)} - \frac{\omega_{pi}^2}{\omega(\omega \pm \Omega_i)} = 0 \tag{4.20}$$

The + sign in (4.20) corresponds to a solution with $E_{1y} = iE_{1x}$, while the negative sign has $E_{1y} = -iE_{1x}$. To see what this means physically we must remember that these solutions are multiplied by $\exp(ikz - i\omega t)$ and that the field is given by the real part of the result. If we choose the origin of t such that $E_{1x} = Re[E_0 \exp(ikz - i\omega t)] = E_0 \cos(kz - \omega t)$, then in the first case $E_{1y} = Re[iE_0 \exp(ikz - i\omega t)] = -E_0 \sin(kz - \omega t)$. The magnitude of the field $(E_{1x}^2 + E_{1y}^2)$ is constant and if we take a fixed value of z and look along the direction of propagation, the electric field vector rotates in a clockwise sense. A wave with this type of behaviour is said to be right circularly polarized. The other wave, with the $-$ sign in (4.20) giving its dispersion relation, has the rotation in the opposite sense and is left circularly polarized. At low frequencies both polarizations tend to the Alfvén wave found previously, the fast and slow modes being indistinguishable for parallel propagation. If we now concentrate on the right circularly polarized wave (with the + sign in (4.20)) we see that, since Ω_e is negative, $k \to \infty$ as $\omega \to |\Omega_e|$. When ω is just above $|\Omega_e|$, n^2 is negative, so that no wave propagates in this frequency range. However, for frequencies much greater than the electron plasma or cyclotron frequencies the solution of (4.20) is approximately $n^2 = 1$, that is $\omega^2 = k^2c^2$, the dispersion relation of the vacuum electromagnetic wave. At some frequency above $|\Omega_e|$ the solution for n^2 must, therefore, pass through zero and become positive again. Putting $n^2 = 0$ in the dispersion relation we find that the resulting equation reduces to

$$\omega^2[(\omega + \Omega_e)(\omega + \Omega_i) - \omega_p^2] = 0$$

which has only one positive root, namely

$$\begin{aligned} \omega = \omega_1 &= -\frac{\Omega_e + \Omega_i}{2} + \left\{\frac{(\Omega_e + \Omega_i)^2}{4} + \omega_p^2\right\}^{1/2} \\ &\approx \left|\frac{\Omega_e}{2}\right| + \left(\frac{\Omega_e^2}{4} + \omega_p^2\right)^{1/2} \\ &> |\Omega_e| \end{aligned} \tag{4.21}$$

Above this frequency the wave propagates once more.

The dispersion curves for the right circularly polarized mode, showing the relation between ω and k, are as shown in Figure 4.1. The continuation of the Alfvén wave above the ion cyclotron frequency gives what is called a whistler, for the following reason. In the earth's magnetic field such modes can

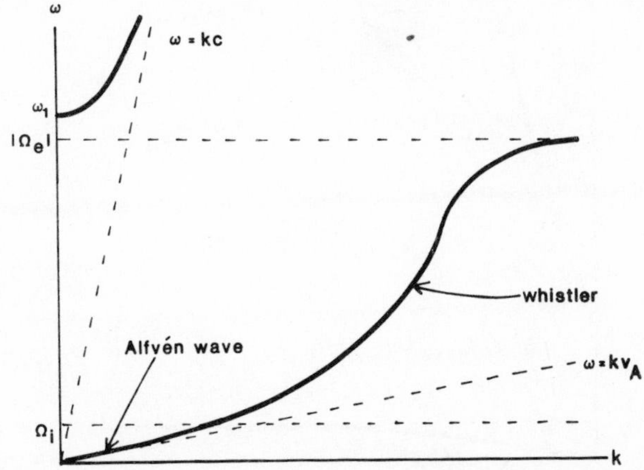

Figure 4.1 Dispersion curves for the right circularly polarized wave propagating parallel to the magnetic field.

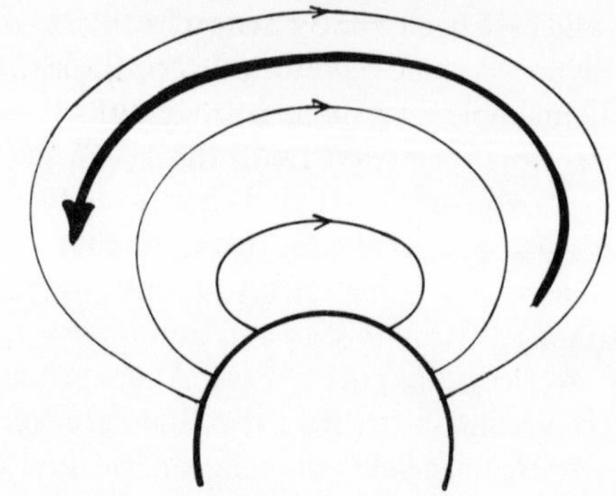

Figure 4.2 Path of a whistler wave in the Earth's dipole field.

propagate at audio frequencies. A lightning flash in one hemisphere produces a pulse which propagates along the Earth's dipole field as shown in Figure 4.2, being channelled back to Earth at the corresponding point in the other hemisphere. A more detailed consideration of the dispersion curves shows that, as drawn in Figure 4.1, the curve is concave up in the whistler region, so that both phase and group velocities increase with frequency. The result is that the initial short pulse, containing a wide band of frequencies is spread out during its passage through the ionosphere, with the higher frequencies arriving first. With suitable equipment the pulse can be heard as a whistle, descending in pitch. This effect was noticed in the early years of this century, but its explanation was not found until much later. The shape of the pulse and the way the frequency varies with time within it can yield a considerable amount of information about the plasma through which it has passed and whistlers have received a great deal of attention with the object of investigating the nature of the Earth's ionosphere.

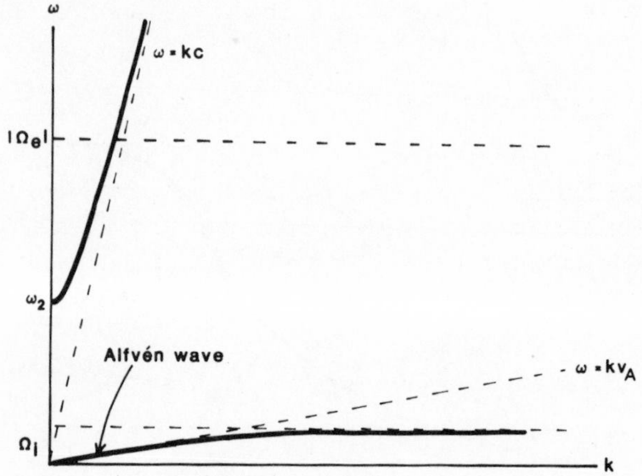

Figure 4.3 Dispersion curves for the left circularly polarized wave propagating parallel to the magnetic field.

For the left circularly polarized wave similar considerations give dispersion curves as shown in Figure 4.3, where now k goes to infinity at the ion cyclotron frequency and the frequency above which propagation occurs once more is given by

$$\omega_2 \approx -\left|\frac{\Omega_e}{2}\right| + \left(\frac{\Omega_e^2}{4} + \omega_p^2\right)^{1/2}. \tag{4.22}$$

Turning our attention to perpendicular propagation we obtain the dispersion relation

$$\begin{vmatrix} \varepsilon_{11} & \varepsilon_{12} & 0 \\ \varepsilon_{21} & \varepsilon_{22} - n^2 & 0 \\ 0 & 0 & \varepsilon_{33} - n^2 \end{vmatrix} = 0. \tag{4.23}$$

Again the equation factorizes easily and there is an obvious root with $\varepsilon_{33} - n^2 = 0$, or $\omega^2 = \omega_p^2 + k^2c^2$, corresponding to a wave with non-zero E_{1z}. Since the propagation is now in the x-direction this is a transverse wave polarized with its electric field along the direction of the steady magnetic field. Particle motions are along the magnetic field and so are unaffected by it, the result being that the dispersion relation is just the same as for the electromagnetic mode in an unmagnetized plasma. This wave is known as the ordinary (or O) mode.

The other mode, has, in general, a combination of E_{1x} and E_{1y} fields and so is of a mixed longitudinal and transverse nature. Its dispersion relation is

$$n^2 = \varepsilon_{22} - \frac{\varepsilon_{12}\varepsilon_{21}}{\varepsilon_{11}}. \tag{4.24}$$

As with parallel propagation, we can get some idea of the behaviour simply by looking at the frequencies for which n goes to zero or infinity. Taking the infinities first we see that these occur when ε_{11} vanishes, that is

$$1 - \frac{\omega_{pe}^2}{\omega^2 - \Omega_e^2} - \frac{\omega_{pi}^2}{\omega^2 - \Omega_i^2} = 0.$$

This is a quadratic in ω^2 which may be solved in the usual way, but for which approximate solutions may be found even more easily. First we note that if the first two terms are equated to zero we get $\omega^2 = \omega_{pe}^2 + \Omega_e^2$ and that if this is substituted in the third term the result is much smaller than unity, so that this is a good approximation to one of the roots. To find the other we use the fact that the product of the roots is

$$\Omega_e^2\Omega_i^2 + \omega_{pe}^2\Omega_i^2 + \omega_{pi}^2\Omega_e^2 \approx \Omega_e^2\Omega_i^2 + \omega_{pi}^2\Omega_e^2,$$

so that the second root is approximately

$$\omega^2 = \frac{\Omega_e^2\Omega_i^2 + \omega_{pi}^2\Omega_e^2}{\omega_{pe}^2 + \Omega_e^2}$$

The first of these roots is greater than the electron cyclotron or plasma frequencies and is called the upper hybrid frequency, while the second, which lies between the electron and ion cyclotron frequencies, is called the lower hybrid frequency. If we now look for the frequencies at which n goes to zero we find, after a straightforward calculation, that these are $\omega = 0$, ω_1 or ω_2 where ω_1 and ω_2 are as given in equations (4.21) and (4.22). At low frequencies the dispersion relation reduces to that of the Alfvén wave, corresponding to the fast or compressional Alfvén wave discussed previously. The compressional Alfvén wave is so named because it is rather like a sound wave with the restoring force provided by magnetic pressure.

Using this information and the fact that the frequencies defined above satisfy $\omega_{LH} < \omega_2 < \omega_{UH} < \omega_1$ we can deduce that the dispersion curves are as in Figure 4.4. The lowest frequency mode is the compressional Alfvén wave, while the other two branches constitute the extraordinary (or X) mode.

For other angles of propagation the dispersion relation does not factorize in such a simple way, although as we have pointed out previously, there are always two values of n for any given frequency. Only if they are both positive, however, will there be two propagating waves at this frequency. In the lowest frequency range the waves are, as discussed above, usually distinguished by the terms fast and slow or, alternatively, shear and compressional. At higher frequencies they may be called O or X modes, depending on which they go over to at perpendicular propagation. Alternatively, if the propagation is near parallel, it may be more convenient to denote them by right and left. Various points may be noted. From (4.11) we see that, in general, waves at arbitrary θ will involve all three electric field components and will be neither purely longitudinal nor purely transverse. Also, both modes are dependent on the magnetic field and it is only at exactly perpendicular incidence that the O-mode has the same dispersion relation as the electromagnetic wave in an

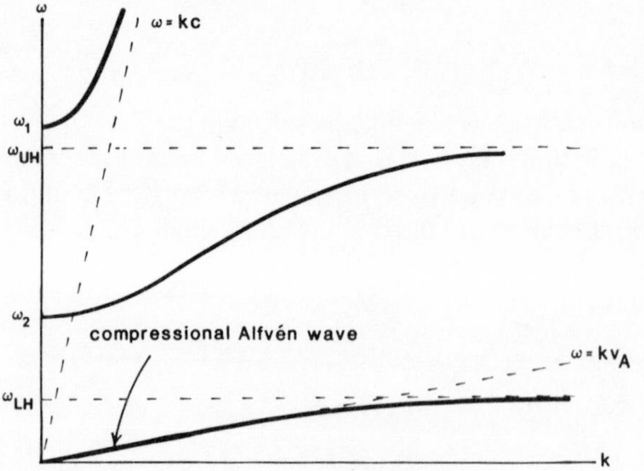

Figure 4.4 Dispersion curves for perpendicular propagation in the X-mode.

unmagnetized plasma. As the propagation direction varies from perpendicular to parallel the X-mode goes to the right circularly polarized wave and the O-mode to the left circularly polarized wave.

At this point it may be useful to say a few words about what happens in a real plasma, where the temperature is not zero. This will be discussed in more detail in the next chapter, but here we simply point out that the main effect is to add extra wave modes. The waves which come out of cold plasma theory retain their essential features in a hot plasma for most frequencies and, since they are the ones which connect smoothly to vacuum electromagnetic wave solutions at the plasma edge, play an important role in the study of wave propagation in a plasma, even if it is at a very high temperature.

4.4 Wave energy density

With a wave propagating in a plasma there is associated the energy of the oscillating electric and magnetic fields, as in a vacuum electromagnetic wave. However, there is also energy associated with the particle motions produced by the oscillating fields, and to obtain sensible and consistent energy relations this must be taken into account.

The total energy density associated with the wave may be calculated most easily by supposing that the wave is driven up by an external agent in such a way that the fields, instead of oscillating sinusoidally, have an exponential growth in amplitude superimposed on their oscillatory behaviour. Thus, instead of going as $e^{-i\omega t}$ the time variation is as $e^{-i\omega t+\gamma t}$, where we will make the further assumption that $\gamma \ll \omega$, so that the growth is over a time which is long compared to the period of oscillation.

The rate of change of the energy density averaged over an oscillation period is

$$\frac{dW}{dt} = \tfrac{1}{2}Re(\boldsymbol{E}^*\cdot\boldsymbol{J}) + \frac{1}{4}\frac{\partial}{\partial t}\left(\varepsilon_0|E|^2 + \frac{1}{\mu_0}|B|^2\right), \tag{4.25}$$

where * denotes the complex conjugate. The first term gives the rate of change of particle energy and the last two the rate of change of electromagnetic field energy. From (4.9) it follows that if ω is real the only component of particle velocity, and hence of current, which is not out of phase with $\boldsymbol{E}$, is perpendicular to $\boldsymbol{E}$, so that the average value of $\boldsymbol{E}\cdot\boldsymbol{J}$ vanishes, as would be expected in a wave of steady amplitude. If, however, the wave is growing, as described above, then the effect is to replace ω with $\omega_1 = \omega + i\gamma$. In terms of the dielectric tensor elements the current is given by

$$J_i = i\omega\varepsilon_0(\delta_{ij} - \varepsilon_{ij}(\omega_1))E_j, \tag{4.26}$$

where we use tensor notation with the usual convention that summation over the repeated index (i.e. j) is to be understood. If we now expand the right-hand side of (4.26) in a Taylor series in γ, and use the fact that γ is small to justify retaining only the terms up to $O(\gamma)$, we obtain

$$J_i = i\omega\varepsilon_0(\delta_{ij} - \varepsilon_{ij}(\omega))E_j - \gamma\varepsilon_0\delta_{ij}E_j + \gamma\varepsilon_0\varepsilon_{ij}(\omega)E_j + \gamma\omega\varepsilon_0\frac{\partial\varepsilon_{ij}(\omega)}{\partial\omega}E_j,$$

which we use in (4.25) to calculate $\mathrm{d}W/\mathrm{d}t$. The contribution from the term independent of γ vanishes since ε_{ij} (which from now on we understand to refer to $\varepsilon_{ij}(\omega)$) is hermitian, implying that $E_i^*\varepsilon_{ij}E_j$ is real. The part of J proportional to γ does, however, give a non-zero contribution, so that

$$\frac{\mathrm{d}W}{\mathrm{d}t} = \tfrac{1}{2}\gamma\varepsilon_0 E_i^*\frac{\partial}{\partial\omega}(\omega\varepsilon_{ij})E_j + \tfrac{1}{2}\gamma\frac{1}{\mu_0}|B|^2, \tag{4.27}$$

in obtaining which we use the relation $(\partial/\partial t)|E|^2 = 2\gamma|E|^2$. Since the time dependence of both sides of (4.27) is as $e^{2\gamma t}$ we may write it in the form

$$\frac{\mathrm{d}W}{\mathrm{d}t} = \frac{\mathrm{d}}{\mathrm{d}t}\left\{\tfrac{1}{4}\varepsilon_0 E_i^*\frac{\partial}{\partial\omega}(\omega\varepsilon_{ij})E_j + \frac{1}{4}\frac{1}{\mu_0}|B|^2\right\}$$

allowing us, finally, to identify the wave energy density as

$$W = \tfrac{1}{4}\varepsilon_0 E_i^*\frac{\partial}{\partial\omega}(\omega\varepsilon_{ij})E_j + \frac{1}{4}\frac{1}{\mu_0}|B|^2. \tag{4.28}$$

In a vacuum $\varepsilon_{ij} = \delta_{ij}$ and the first term reduces to the usual electric field energy density $\tfrac{1}{4}\varepsilon_0|E|^2$ (remember that an extra factor $\tfrac{1}{2}$ has come from the process of averaging over an oscillation period). In a plasma the form (4.28) automatically includes the part of the energy density associated with modifications of the particle motion.

To illustrate this let us consider the simple case of an electromagnetic wave in an unmagnetized plasma, where, according to the above

$$\begin{aligned} W &= \tfrac{1}{4}\varepsilon_0|E|^2\frac{\partial}{\partial\omega}\left[\omega\left(1 - \frac{\omega_p^2}{\omega^2}\right)\right] + \frac{1}{4}\frac{1}{\mu_0}|B|^2 \\ &= \tfrac{1}{4}\varepsilon_0|E|^2 + \frac{1}{4}\frac{1}{\mu_0}|B|^2 + \tfrac{1}{4}\varepsilon_0\frac{\omega_p^2}{\omega^2}|E|^2. \end{aligned}$$

There is, therefore, a term $(\varepsilon_0/4)(\omega_p^2/\omega^2)|E|^2$ in addition to the electromagnetic field energy density. Using

$$i\omega\boldsymbol{v}_{1e} = -\frac{e}{m}\boldsymbol{E},$$

and the corresponding equation for the ion velocity perturbation we see that this extra term is just equal to

$$\tfrac{1}{4}m_e|v_{1e}|^2 + \tfrac{1}{4}m_i|v_{1i}|^2,$$

the average kinetic energy associated with the oscillatory motion of the particles.

It is of interest to note that (4.28) is not necessarily always positive, and that it is possible to have negative energy waves. These do not occur with the cold plasma dielectric tensor which we have calculated, but may occur if the unperturbed plasma contains particle beams or some other source of energy. The physical interpretation of a negative energy wave is that as its amplitude increases the energy density in the system decreases. Clearly the part of the energy density associated with the electromagnetic fields increases with amplitude, but if the particles have some non-equilibrium form of kinetic energy in the unperturbed system it is possible for the excitation of the wave to reduce their average energy and for this effect to outweigh the increase in electromagnetic energy. We shall not pursue this subject here, but simply note that negative energy waves play an important role in many instabilities and recognition of their presence may simplify stability analysis and give a simple physical interpretation of instability mechanisms.

4.5 Waves in an inhomogeneous plasma

In real life, plasmas are not usually the spatially homogeneous systems we have assumed in analysing waves in the first part of this chapter, and we must now tackle the question of what happens in an inhomogeneous plasma, and to what extent the plane wave solutions which we have discovered are useful. Intuitively it might be expected that if the plasma does not vary much over length scales of the order of the wavelength, then the wave would behave locally like the homogeneous plasma solution. Application of this idea leads us to develop the type of approximate treatment of the inhomogeneous plasma problem which is generally known as the WKB approximation (after Wentzel, Kramers and Brillouin who introduced it into quantum mechanics).

To see the basic idea of this method we will apply it initially to a simpler problem than the complicated wave equations which are obtained for a plasma. Let us instead look at the equation

$$\frac{\partial^2 a}{\partial x^2} = f(x)\frac{\partial^2 a}{\partial t^2}, \tag{4.29}$$

that is a standard one-dimensional wave equation with an x-dependent wave speed. We can represent the time dependence as a superposition of Fourier components going as $e^{-i\omega t}$, so we take

$$a(x, t) = a_0(x)e^{-i\omega t},$$

with a_0 obeying the ordinary differential equation

$$\frac{d^2 a_0}{dx^2} + \omega^2 f(x) a_0 = 0. \tag{4.30}$$

We consider first a simple extension of the $e^{\pm ikx}$ dependence which would be a solution for constant f, and look at the extent to which the equation is satisfied by $a_0 = A\exp(\pm\int_0^x k(x')\,dx')$ with $k^2 = \omega^2 f(x)$. The second derivative of this expression is $-k^2 a_0 \pm i(dk/dx)a_0$, so that there is an extra term $\pm i(dk/dx)a_0$ left over when we substitute into (4.30). However, in line with the basic idea of the method, the gradient of $f(x)$, and hence of k, is to be assumed small, in the sense that $|df/dx| \ll kf$, so that this left-over term is small compared with the two which cancel. What we have shown, then, is that a solution going as $\exp(\pm i\int_0^x k(x')\,dx')$, with $k(x)$ the local value of the wavenumber, is a reasonable approximation. The same holds in the more complicated plasma problem, so that the homogeneous plasma solutions which we have analysed in some detail provide a great deal of information on the behaviour to be expected when gradients are present. We should point out here that the choice of the lower limit of the integral to be zero is not important and changing it just multiplies the solution by a constant.

A better approximation, and one which reveals the limitations of the technique, may be obtained by assuming that the amplitude of the solution is x-dependent, so that we take

$$a_0 = A(x)\exp\left(\pm i\int_0^x k(x')\,dx'\right). \tag{4.31}$$

If this is substituted in the equation and if we neglect the second derivative of A we obtain

$$2kA' + k'A = 0,$$

which is satisfied if A is proportional to $|k|^{-1/2}$. The neglect of the second derivative of A is again based on the assumption that k is a slowly varying smooth function. It is clear, however, from the fact that A goes as $|k|^{-1/2}$, that the method breaks down and A is not slowly varying in the neighbourhood of a point where $k = 0$. Small k corresponds to long wavelength, so that if k is too small the assumptions underlying the method break down.

We shall now turn to the problem of waves in a cold inhomogeneous plasma. A basic complication over our simple example is of course that the problem is three-dimensional. Rather than deal with a general three-dimensional geometry, however, we shall consider the case where the plasma has gradients along one direction only, this often being a reasonable approximation to practical situations if a limited region is being considered. Wave propagation in a tokamak, for example, may be understood to a considerable extent by looking at a plasma where the curvature of the magnetic field is neglected and the gradients are taken to be perpendicular to the field direction. To illustrate wave behaviour in an inhomogeneous plasma we shall, in fact, consider this last geometry so that, if we use the coordinate system we had before, the magnetic field is along the z-direction and the plasma gradients are in the x-direction. Since there is no variation of the unperturbed system with z the perturbations may simply be taken to vary as $e^{ik_z z - i\omega t}$. For the x-dependence we take, following our earlier procedure, $\exp(i\int_0^x k_x(x')\,dx')$, and as before we see that if k_x satisfies the homogeneous plasma dispersion relation at each point then the equations will be satisfied to within terms involving dk_x/dx. By demanding that the equations be satisfied to higher order we may see how the wave amplitude varies as it propagates.

An equivalent and physically more transparent way of calculating the variation of wave amplitude as the wave propagates is to note that in a non-dissipative system such as we are considering energy must be conserved, implying that the divergence of the energy flux must be zero. This latter quantity is given by the group velocity $\partial\omega/\partial\boldsymbol{k}$ times the energy density calculated in the previous section, so that for the geometry we are considering we obtain

$$\frac{d}{dx}\left[\frac{\partial\omega}{\partial k_x}W\right] = 0, \tag{4.32}$$

with W given by (4.28). The relation between $\boldsymbol{E}$ and $\boldsymbol{B}$ and amongst the components of $\boldsymbol{E}$ is just that appropriate to the solution of the local homogeneous wave problem, so that the single equation (4.32) can yield the variation on the long length scale.

As in our simple example there are points at which the approximation breaks down, most notably those at which k_x goes to zero or infinity, known as cut-offs and resonances respectively. Near each of these $\partial\omega/\partial k_x \to 0$ and it is clear that the amplitude calculated from (4.32) is no longer slowly varying. The other circumstance under which the approximation fails is when there are two almost coincident roots of k_x, in which case $\partial\omega/\partial k_x$ is again small.

To summarize our picture of wave propagation so far, we have seen that in a homogeneous plasma there are various possible modes which may propagate. In an inhomogeneous plasma these also propagate independently with slowly varying wavelength and amplitude, prescribed by the local dispersion relation and energy conservation, until they reach one of the regions in which the WKB approximation fails. In the next section we shall discuss what happens in these regions, something which is obviously necessary before our description is of much value, then use our accumulated knowledge to look at some problems of practical interest in section 4.7.

4.6 Cut-off and resonance

The dispersion relation to be solved for k_x in the geometry of the previous section is, in fact, a quadratic in k_x, so that in a region where a wave propagates there is at least one positive root for k_x^2, giving two real values of k of opposite sign which correspond to the two directions of propagation of the wave along the x-axis. At a cut-off k_x^2 goes from positive to negative, so that beyond the cut-off the wave is evanescent. In order to see what happens at the cut-off we must return to the exact equation for the wave, the essential features of which are reproduced by the equation

$$\frac{d^2\phi}{dx^2} + x\phi = 0. \tag{4.33}$$

The local wavenumber corresponding to the WKB approximations for this equation is given by $k^2 = x$ and so solutions for $x > 0$ go as $\exp(\pm \frac{2}{3}ix^{3/2})$, representing propagating waves going in the two directions, and for $x < 0$ go as $\exp(\pm \frac{2}{3}|x|^{3/2})$. For a physically realistic solution we would wish the wave amplitude to decay in the region $x < 0$, so our problem is to find a solution with this property. Imposing this condition is equivalent to choosing one of the two linearly independent solutions of the equation (4.33) and fixes the behaviour in the region $x > 0$. As to the solution of equation (4.33) we are fortunate in that it is one of the standard equations of mathematical physics, Airy's equation, the solutions of which are well-documented. The solution with the property required is generally denoted by $Ai(-x)$, given by

$$Ai(-x) = \frac{1}{\pi}\int_0^\infty \cos(-xt + \tfrac{1}{3}t^3)\,dt. \tag{4.34}$$

For our purposes we need only note that for large positive x

$$Ai(-x) \sim \frac{1}{\sqrt{\pi}} x^{-1/4} \sin\left(\tfrac{3}{2}x^{3/2} + \frac{\pi}{4}\right) \tag{4.35}$$

which, if the sine is expressed in terms of exponentials, represents a superposition of equal amplitudes of right- and left-propagating waves, implying that a wave incident on the cut-off at $x = 0$ is completely reflected. The behaviour of the Airy function Ai is illustrated in Figure 4.5, showing essentially a standing wave pattern produced by the interference of the incoming and outgoing waves.

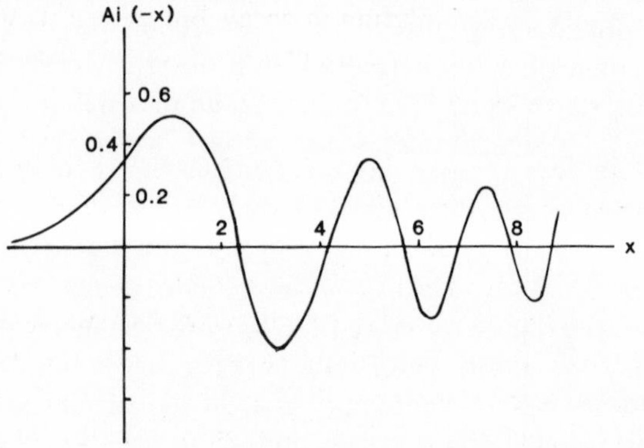

Figure 4.5 Behaviour of the solution of (4.33).

In general, when k_x^2 passes through zero at a cut-off, its behaviour around the cut-off may be approximated by a linear dependence on x, and the wave amplitude has the behaviour illustrated in Figure 4.5, apart from a rescaling of the x-axis. We arrive, therefore, at the important conclusion that at a cut-off a wave is totally reflected.

Similarly, in the neighbourhood of a resonance it may be supposed, in general, that k_x^2 varies as $1/x$, and that we can infer the behaviour to be expected by examining the equation

$$\frac{d^2\phi}{dx^2} + \frac{1}{x}\phi = 0, \tag{4.36}$$

the solutions of which are $x^{1/2}\psi(2x^{1/2})$ with $\psi(y)$ satisfying

$$\frac{d^2\psi}{dy^2} + \frac{1}{y}\frac{d\psi}{dy} + \left(1 - \frac{1}{y^2}\right)\psi = 0, \tag{4.37}$$

that is Bessel's equations of order one. Thus

$$\phi = c_1 x^{1/2} J_1(2x^{1/2}) + c_2 x^{1/2} Y_1(2x^{1/2}),$$

with J_1 and Y_1 the Bessel functions of the first and second kind. For large $|x|$,

$$J_1(y) \sim \left(\frac{2}{\pi y}\right)^{1/2} \cos\left(y - \tfrac{3}{4}\pi\right)$$

$$Y_1(y) \sim \left(\frac{2}{\pi y}\right) \sin\left(y - \tfrac{3}{4}\pi\right),$$

so that for large $|x|$

$$\phi \sim \frac{x^{1/2}}{2}\left(\frac{1}{\pi x^{1/2}}\right)^{1/2} \left[(c_1 + c_2)\exp\left(2ix^{1/2} - \tfrac{3}{4}\pi\right)\right.$$
$$\left. + i(c_1 - c_2)\exp\left(-2ix^{1/2} - \tfrac{3}{4}\pi\right)\right].$$

For negative x we wish the solution to decay, but there is a difficulty because which of the two terms in the square bracket above decays depends on how we choose the argument of x. Another problem is that Y_1 is singular at the origin, but its contribution is needed to give an acceptable solution for large negative x. One way of resolving these difficulties is to introduce a small amount of damping into the system, for example by adding a term $+\gamma_s v_{1s}$ to the left-hand side of (4.5). The effect of this is to move the singularity in k_x^2 off the real axis and into the upper half complex plane. We therefore take the cut in the complex plane, which is necessary to make x single-valued, to be in the upper half plane. Thus the argument of x on the real axis, if taken to be zero for positive values, goes to $-\pi$ for negative values. For negative x, $x^{1/2} = -i|x|^{1/2}$ and so a decaying solution is given by the second term in square brackets. The first must be made to vanish by taking $c_1 = -c_2$. For positive x, the solution goes as $e^{-2ix^{1/2}-i\omega t}$, if we put in the time-dependence, and represents a wave travelling towards $x = 0$. The conclusion is therefore that the wave is totally absorbed at a resonance. The absorption is provided by the damping, but is complete, regardless of the magnitude of the damping coefficients. What happens is that if the damping is small the fields become very large in the neighbourhood of the resonance and there is a large amount of absorption in a very small region. With larger damping the peak amplitudes are reduced, but damping takes place over a wider region. The problem can also be treated without introducing damping, but only by abandoning the assumption that the time variation is sinusoidal and treating the problem as an initial value problem instead. In this case the wave amplitude around the resonance grows in time and the energy of the incoming wave goes into the establishment of this ever-growing spike in the amplitude. The above is a brief and incomplete account of a rather complicated subject, but the essential conclusion which should be remembered is that wave energy is absorbed at a resonance. A somewhat similar phenomenon is studied in more detail in section 4.8.

Sometimes the physical mechanism leading to absorption is clear. For example the right-hand circularly polarized wave propagating along the magnetic field has a resonance when its frequency equals the electron cyclotron frequency. In this case the electric field vector is rotating in the same sense as the electrons so that an electron experiences a steady electric field which tends to accelerate it. Other cases, like the hybrid resonances for perpendicular propagation, do not have such an obvious interpretation.

The other circumstance in which the WKB approximation may break down is when two solutions of the dispersion relation are almost coincident. In the neighbourhood of such a point two modes which are quite separate over most of the plasma have almost identical characteristics and the typical result is a 'mode conversion', in which energy incident entirely in one mode emerges partitioned between the two modes. Such processes are very important in understanding the mechanisms through which radio-frequency waves may heat magnetically confined plasmas, but their theory is beyond the scope of our present discussion.

4.7 Propagation of electron cyclotron waves in a tokamak

In order to heat the plasma in a tokamak to the temperature required for fusion it is necessary to use some means in addition to the resistive heating arising from the current in the plasma. One such means, which has been the subject of intensive investigation in recent years, is absorption of energy from higher-power radio-frequency waves directed into the plasma from an external source. Various frequency ranges give effective absorption and here, in order to give an important practical illustration of the techniques developed

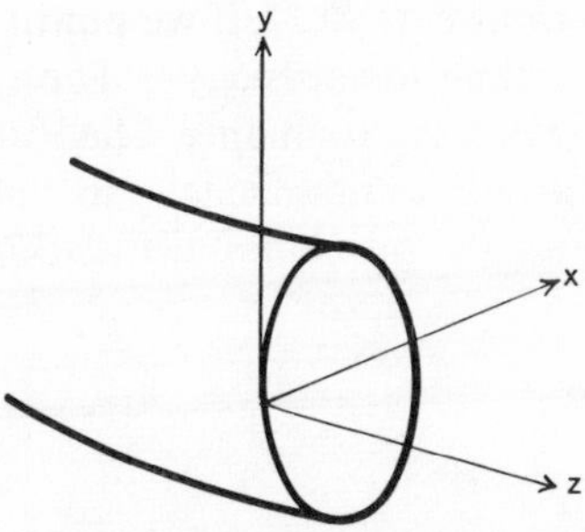

Figure 4.6 Alignment of the axes in the tokamak geometry.

in the earlier part of this chapter, we shall discuss one of these, in which the wave frequency coincides with the electron cyclotron frequency somewhere near the centre of the plasma.

For the magnetic field typical of a large tokamak this implies a wavelength of the order of 1 cm, and so the WKB approximation may be expected to be a good one in such a machine. If the wave is incident in the mid-plane of the torus, its propagation is, to a reasonable approximation, described by a model with the geometry of section 4.5, when the axes are aligned with respect to the torus as illustrated in Figure 4.6. The magnetic field is taken to be in the z-direction and to be uniform, so we are neglecting the field curvature and the effect of the poloidal field. The variation of density and magnetic field along the x-axis is as illustrated in Figure 4.7.

The parameters of a tokamak are generally such that the electron plasma frequencies and cyclotron frequencies are of the same order of magnitude near the centre of the plasma, so that an electron cyclotron wave has, everywhere, a frequency much greater than the ion plasma and cyclotron frequencies. This means that we can neglect the ion contributions to the dispersion relation.

As before, it is convenient to work with $\boldsymbol{n} = \omega \boldsymbol{k}/c$, rather than $\boldsymbol{k}$ itself, and our problem requires that we find n_x, for a fixed value of n_z. The y-component

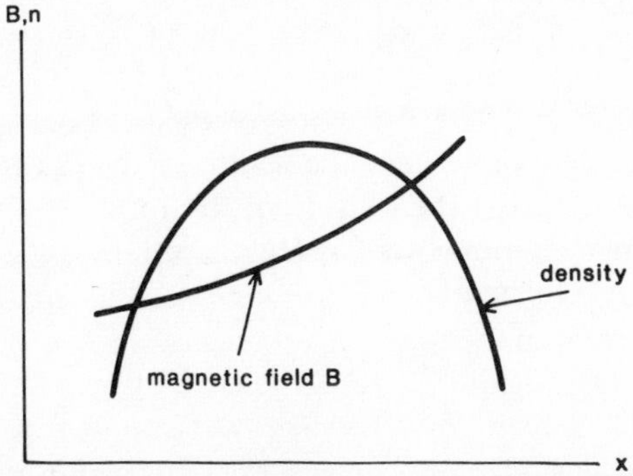

Figure 4.7 Variation of magnetic field and density across the tokamak.

of $\boldsymbol{n}$ is taken to be zero, though it should be noted that, since the x-axis is fixed by the direction of the inhomogeneity, this no longer gives the most general case of wave propagation. Including n_y, which is constant if plasma properties are independent of y, complicates the algebra of the dispersion relation, but does not affect the principles of the method. To find n_x we go back to (4.16) and write it in the form

$$\begin{vmatrix} \varepsilon_{11}-n_z^2 & \varepsilon_{12} & n_x n_z \\ \varepsilon_{21} & \varepsilon_{22}-n_x^2-n_z^2 & 0 \\ n_x n_z & 0 & \varepsilon_{33}-n_x^2 \end{vmatrix} = 0, \tag{4.38}$$

which turns out, on expanding the determinant, to give a quadratic in n_x^2,

$$\varepsilon_{11} n_x^4 - b n_x^2 + c = 0 \tag{4.39}$$

with

$$b = (\varepsilon_{11}+\varepsilon_{33})(\varepsilon_{11}-n_z^2) - \varepsilon_{12}\varepsilon_{21}$$
$$c = \varepsilon_{33}[(\varepsilon_{11}-n_z^2)^2 - \varepsilon_{12}\varepsilon_{21}].$$

Let us begin by looking at incidence exactly perpendicular to the magnetic field, that is, with $n_z = 0$. Then, as we have seen before the determinant (4.38) factorizes in an obvious way into the ordinary mode with $n_x^2 = \varepsilon_{33} = 1-(\omega_{pe}^2/\omega^2)$ and the extraordinary mode, with $n_x^2 = \varepsilon_{22} - \varepsilon_{12}\varepsilon_{21}/\varepsilon_{11}$. The O-mode propagates if $\omega_{pe}^2 \leqslant \omega^2$, regardless of the magnetic field, so that according to our theory we would expect it to be reflected from the plasma if the density in the centre were large enough, and otherwise to pass straight through. If the effects of plasma temperature are taken into account it is found that there should be strong absorption in a large tokamak where the frequency equals the electron cyclotron frequency. An effective heating mechanism is obtained by tuning the incoming wave to the cyclotron frequency near the centre of the plasma column, provided that the density is not so high that the wave is reflected before it reaches the centre.

Turning to the X-mode, we recall that it has a resonance, the upper hybrid resonance, when $\omega^2 = \omega_{pe}^2 + \Omega_e^2$ and cut-offs when

$$\omega = \pm\left|\frac{\Omega_e}{2}\right| + \left(\frac{\Omega_e^2}{4} + \omega_p^2\right)^{1/2}.$$

If, again $\omega^2 = \Omega_e^2$ somewhere near the centre of the plasma column, then since Ω_e^2 decreases towards the outside of the torus we will find that the upper hybrid resonance occurs in this region. Still nearer the plasma edge, we come to the cut-off where

$$\omega = \left|\frac{\Omega_e}{2}\right| + \left(\frac{\Omega_e^2}{4} + \omega_p^2\right)^{1/2}.$$

The location of the cut-off and resonance are indicated in Figure 4.8. To the left of the cut-off the wave propagates as the highest-frequency branch in

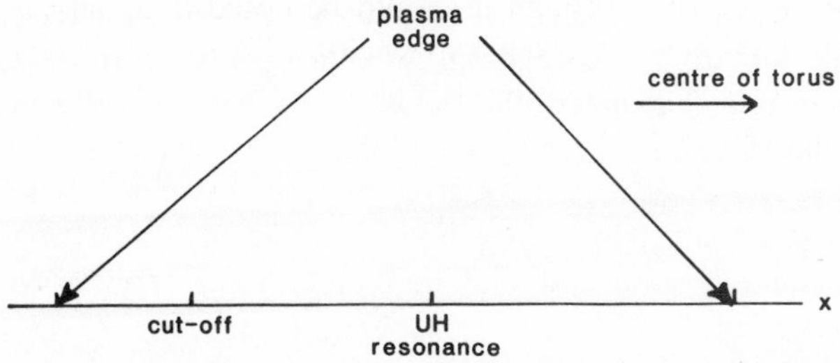

Figure 4.8 Positions of cut-off and resonance for the X-mode at perpendicular incidence.

Figure 4.4, then between the cut-off and resonance the wave does not propagate, while to the right of the resonance the wave propagates as the middle branch in Figure 4.4.

We see that the behaviour to be expected of the X-mode is that it is reflected at the cut-off if it comes from the outside of the torus, while if it comes from the inside it is absorbed at the upper hybrid resonance provided that the density is not so large that the other cut-off at ω_2 exists in the plasma. Again finite temperature effects may produce absorption where $\omega = |\Omega_e|$, but this point, lying to the right of the upper hybrid resonance, is accessible only from the inside of the torus. Thus, if we wish to produce heating using the X-mode in the electron cyclotron frequency range we must have it incident from the inside of the torus, which is technologically less convenient than having it incident from the outside. For both the O- and X-modes there is a density limit above which the centre of the plasma is not accessible to the wave.

Away from exactly perpendicular incidence n_x^2 is easily found from (4.39), even though the quadratic does not factorize in quite so obvious a fashion. When ε_{11} goes to zero, one root tends to infinity while the other tends to a finite value. Thus one of the modes still has a resonance at the upper hybrid frequency and is called the extraordinary mode, since it goes smoothly into the extraordinary mode at normal incidence. The other mode, the ordinary mode, has no resonance. A cut-off, where $n_x = 0$, occurs when the constant coefficient c of (4.39) goes to zero. One such cut-off corresponds to $\varepsilon_{33} = 1 - \omega_p^2/\omega^2 = 0$, and is just the cut-off of the O-mode, whose position is independent of n_z. The cut-offs of the X-mode do change position as n_z varies, but the overall picture found for normal incidence remains the same.

The fact that in the geometry being considered here the position of a resonance is independent of n_z may be explained quite simply by noting that if n_z is fixed, then as n_x tends to infinity the wave vector tends to become perpendicular to the magnetic field, so the resonance is just the same as the resonance for perpendicular incidence. The direction of propagation of the wave is not constant, and solutions of the dispersion relation for constant θ, which are sometimes presented, may not be directly applicable to our present problem.

The dependence of all quantities on one spatial dimension is clearly a rather rough approximation to a tokamak or other confinement device. In a full-three-dimensional geometry, the cut-offs and resonances generally lie on curved surfaces and 'ray-tracing' computer programs are used to follow the flow of wave energy. We shall not go into this in any detail, but simply point out that our conclusions on the direction from which heating regions are accessible remain valid, and that one-dimensional models do give useful information.

Before leaving this subject we shall mention briefly some of the other frequency regimes which are of interest for radio-frequency heating of tokamaks. Typical frequencies for electron cyclotron heating are of the order of 20–100 GHz (1 GHz = 10^9 Hz), and going down in frequency, the next useful range is that at which the lower hybrid resonance provides the absorption, at typical frequencies of a few GHz. Then there is absorption at the ion cyclotron frequency or its low harmonics, at about 10–100 MHz, and finally heating by excitation of Alfvén waves in the system, at frequencies of a few MHz. In all of these regimes considerations like those above can be useful in giving information about the propagation of waves and the accessibility of the plasma to them.

Experiments have been carried out on all of these, but mostly at the ion cyclotron and lower hybrid frequencies, for which a number of laboratories have demonstrated effective absorption of the wave energy resulting in plasma heating. At the lower frequencies an antenna structure inside the toroidal vacuum vessel is needed, while for higher frequencies a waveguide may be used to introduce the waves into the vessel. At the high frequencies of electron cyclotron waves, waveguides are very easily used to transport energy from the source to the plasma, and the main reason why this is somewhat less developed is the difficulty of obtaining high-power sources at the very high frequencies needed. Recently there has been rapid development of devices known as gyrotrons, which satisfy the requirements for electron cyclotron heating, and experiments have begun to show that this method is also capable of heating a plasma efficiently.

4.8 Tunnelling and resonant absorption

In discussing the behaviour of the X-mode incident from the outside of a tokamak we have said that it is reflected on reaching the cut-off. However, the behaviour of the wave amplitude near the cut-off is as illustrated in Figure 4.5, with a gradual decay into the region beyond the cut-off, rather than an abrupt drop to zero. If the distance between the cut-off and the upper hybrid resonance is small enough, some of the incident wave energy may tunnel through the region of evanescence and propagate into the region beyond the upper hybrid resonance, a phenomenon of exactly the same type as quantum mechanical tunnelling through a potential barrier. For this to be important the thickness of the non-propagating region must not be much more than the wavelength, and at electron cyclotron frequencies it is significant only in very small devices. In the lower frequency ranges it may be more significant and there is, indeed, in some of the lower-frequency heating schemes, a non-propagating layer at the plasma edge, through which the wave must tunnel from the antenna.

This phenomenon is not, of course, described by the WKB approximation and it is again necessary to return to more exact equations in order to find the fraction of the incident energy which tunnels through the barrier. In what follows we shall examine in detail one of the simpler problems of this type, involving an unmagnetized plasma. In so doing we shall uncover an effect known as resonant absorption, which is very important in the interaction of laser light with a plasma.

The plasma we wish to examine is one with a density gradient along the x-direction on which light of angular frequency ω is incident from the low density side, as illustrated in Figure 4.9.

The density is assumed to reach a value such that at some value of x, defining what is called the critical surface, the plasma frequency becomes equal to the incident frequency. If laser light is incident on a solid target the density near the centre of the target is much greater than the density at the critical surface (the critical density), and so Figure 4.9. represents the profile in the outer regions of the target plasma.

If light is incident obliquely to the critical surface, in the x–y plane, then the local dispersion relation is

$$\omega^2 = \omega_p^2 + k^2c^2 = \omega_p^2 + (k_x^2 + k_y^2)c^2. \tag{4.40}$$

As the light propagates k_y remains constant and

$$k_x^2 = \frac{\omega^2 - \omega_p^2}{c^2} - k_y^2$$

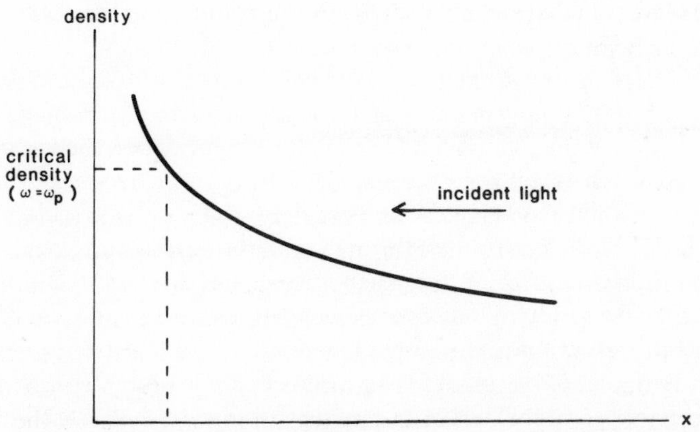

Figure 4.9 Plasma density profile for resonant absorption problem.

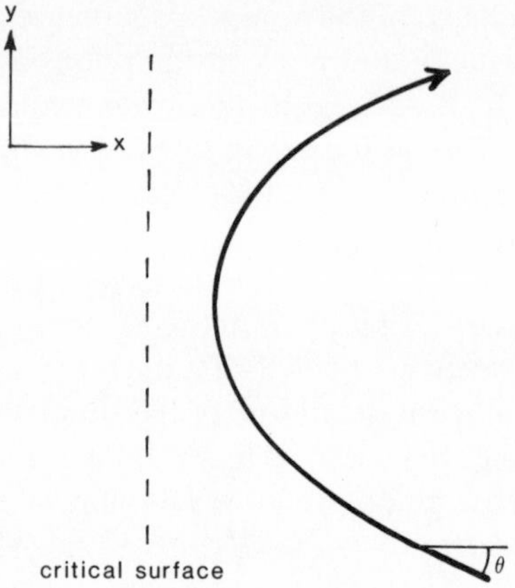

Figure 4.10 Path of a light ray in the WKB approximation.

varies. If θ is the angle of incidence of the light in the vacuum region outside the plasma, $k_y = k_0 \sin\theta = (\omega/c)\sin\theta$, where k_0 is the vacuum wavenumber. Thus $k_x^2 = (\omega^2 \cos^2\theta - \omega_p^2)/c^2$ giving a cut-off where $\omega = \omega_p \cos\theta$ so that in the WKB approximation we expect the light rays to follow a path like that shown in Figure 4.10, with total reflection at, or on the low-density side of, the critical surface, the point of reflection moving outwards from the critical surface as θ increases.

If, however, we look at the exact equations for the fields in this geometry we find that the above picture is not correct for the component of the incoming wave polarized with its electric field in the x–y plane, but is valid only if the wave is polarized with its electric field perpendicular to the plane of incidence. We shall begin by explaining in physical terms why this should be so, then look at the mathematics of the problem.

The key to understanding the process is to remember that in an unmagnetized plasma there is another wave mode in addition to the electromagnetic wave, namely the plasma wave. As we have seen this oscillates at the frequency ω_p, regardless of its wavelength, so it will be localized around the critical surface in the inhomogeneous system. If the incident electromagnetic wave is polarized with the electric field in the plane of incidence, then at its turning point there will be a component along the density gradient, which will produce a charge imbalance as the electrons oscillate while the ions remain virtually at rest. If this oscillation tunnels through to the critical surface it will constitute a resonant driving force on the plasma wave and drive up the amplitude of the latter. The result is that energy is abstracted from the incident wave, reducing the reflection coefficient below one, and is fed into the plasma wave. If a steady state is to be achieved there must be some sink of energy to prevent indefinite growth of this wave, and in view of this we shall introduce a small amount of damping into the system, writing the linearized equation for the electron velocity perturbation as

$$\frac{\partial \boldsymbol{v}_e}{\partial t} + \nu \boldsymbol{v}_e = -\frac{e}{m}\boldsymbol{E}$$

or, taking the perturbations to go as $e^{ik_y y - i\omega t}$ times a function of x,

$$(-i\omega + \nu)\boldsymbol{v}_e = -\frac{e}{m}\boldsymbol{E}. \tag{4.41}$$

If $\boldsymbol{E}$ is in the x–y plane, then $\boldsymbol{B}$ is in the z-direction, and taking the z-component of $\boldsymbol{\nabla} \times \boldsymbol{E} = -\partial \boldsymbol{B}/\partial t$ we get

$$ik_y E_x - \frac{\mathrm{d}E_y}{\mathrm{d}x} = i\omega B. \tag{4.42}$$

Finally we complete the set of equations by taking the x and y components of

$$\boldsymbol{\nabla} \times \boldsymbol{B} = -\mu_0 n_0 e \boldsymbol{v}_e - \frac{i\omega}{c^2}\boldsymbol{E},$$

namely

$$ik_y B = -\mu_0 n_0 e v_{ex} - \frac{i\omega}{c^2} E_x \tag{4.43}$$

$$-\frac{\mathrm{d}B}{\mathrm{d}x} = -\mu_0 n_0 e v_{ey} - i\frac{\omega}{c^2} E_y. \tag{4.44}$$

From (4.41), (4.43) and (4.44) we may find E_x and E_y in terms of B and $\mathrm{d}B/\mathrm{d}x$ and, substituting this into (4.42), obtain a differential equation for B (remembering that n_0 depends on x),

$$\frac{\mathrm{d}^2 B}{\mathrm{d}x^2} - \frac{1}{\varepsilon}\frac{\mathrm{d}\varepsilon}{\mathrm{d}x}\frac{\mathrm{d}B}{\mathrm{d}x} + k_0^2(\varepsilon - \sin^2\theta)B = 0, \tag{4.45}$$

where $\varepsilon = 1 - \omega_p^2/[\omega(\omega + i\nu)]$. We shall assume that the profile around the critical density may reasonably be represented by a linear density gradient so that $\omega_p^2 = \omega^2[1 - (x/a)]$ with the origin taken to be at the critical surface. Then, assuming $\nu \ll \omega$, $\varepsilon \simeq (x/a) - (i\nu/\omega)$ near the critical surface, and the equation is

$$\frac{\mathrm{d}^2 B}{\mathrm{d}x^2} - \frac{1}{x - i\nu a/\omega}\frac{\mathrm{d}B}{\mathrm{d}x} + k_0^2\left[\frac{x}{a} - \sin^2\theta\right]B = 0,$$

where we have neglected the imaginary part of ε in the coefficient of B, but must keep it in the coefficient of $\mathrm{d}B/\mathrm{d}x$ in order to avoid having a singularity on the real axis. It is useful to make the change of variable $x = (ak_0^2)^{1/3}\xi$, so that the equation becomes

$$\frac{\mathrm{d}^2 B}{\mathrm{d}\xi^2} - \frac{1}{\xi - i\gamma}\frac{\mathrm{d}B}{\mathrm{d}\xi} + (\xi - q)B = 0, \tag{4.46}$$

$$\text{with } q=(k_0a)^{2/3}\sin^2\theta \quad \text{and} \quad \gamma=\frac{va}{\omega(ak_0^2)^{1/3}}.$$

The presence of γ is only important near $\xi=0$, and if we look for the leading order terms in an expansion of B around the singularity at $\xi-i\gamma=0$ we find that B goes as $1+(q/2)(\xi-i\gamma)^2\log(\xi-i\gamma)$, while E_x and E_y go as $1/(\xi-i\gamma)$ and $\log(\xi-i\gamma)$ respectively. The presence of the damping prevents the fields from being singular on the real axis. The fundamental quantity of interest is the absorption coefficient of the incident radiation, which is found from the solution for large positive x. This consists of a superposition of incoming and outgoing waves, whose ratio is fixed by the boundary condition which must be satisfied for large negative x, namely that the wave should decay in this region. The important role played by the damping is to determine the way in which the solution is continued through the region around $\xi=0$. With no damping the solution here depends on $\log\xi$ and it is not clear whether, if the argument of ξ is zero for ξ positive, it should be π or $-\pi$ for ξ negative. Changing this choice changes $\log\xi$ by $2\pi i$ to the left of the singularity and so makes a considerable difference to the overall solution. With damping included the singularity is moved into the upper half plane and to get a continuous solution along the real axis the cut in the complex plane must be taken upwards as shown in Figure 4.11.

The argument of ξ must then vary continuously in the lower half plane, and is therefore $-\pi$ on the negative real axis. So long as the damping is small it plays no important role in the solution other than to resolve this ambiguity and the reflection coefficient is independent of it, depending only on the single parameter q. Equation (4.46) can be solved numerically, or an analytic approximation obtained by piecing together approximations valid in different regions, the absorbed fraction of the incident energy being found to be as illustrated in Figure 4.12.

The shape of the curve in Figure 4.12 can be understood as follows. At

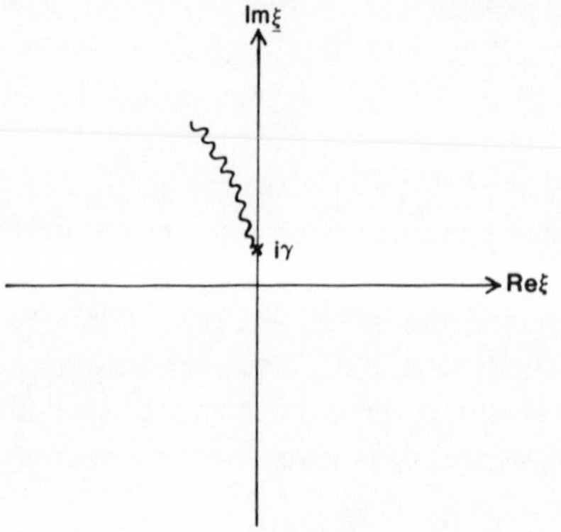

Figure 4.11 Singularity and branch cut in the complex plane.

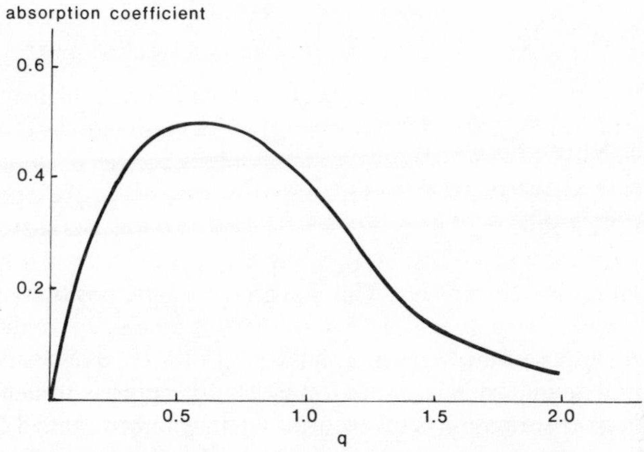

Figure 4.12 Absorption coefficient for resonant absorption.

normal incidence there is no field component along the density gradient, and so no way of coupling the electromagnetic wave to the plasma wave. As the angle increases, the component along the gradient increases and absorption increases. For large angles, however, the turning point moves away from the critical surface and the increase in the tunnelling length eventually causes the absorption to decrease.

If the density length scale is long compared to the wavelength, as might be the case in the ionosphere, for example, resonant absorption is only effective in a very narrow range of angles close to perpendicular. In a laser-irradiated target, however, the high electromagnetic field intensity near the critical surface produces a steepening of the density profile, the mechanism for which (the ponderomotive force) is discussed in Chapter 6. The result is that resonant absorption may be effective over a wide range of angles and plays an important role in laser–plasma interactions. Experiments have verified that absorption of laser light is dependent on the polarization of the wave and have shown an angular dependence which is in reasonable agreement with that of Figure 4.12. The situation is, of course, complicated by other absorption mechanisms and the fact that the geometry of the critical surface is more complicated than we have assumed.

We have seen that the fraction of the incident energy which is absorbed is independent of the damping. Since this may appear rather surprising at first sight we shall examine the underlying physical behaviour in more detail. For this purpose it is sufficient to recognize that the electrons behave, in essence, like harmonic oscillators whose natural frequency is the plasma frequency. When driven by an external force per unit mass $F_0 e^{-i\omega t}$, the response of such an oscillator is given by the equation

$$\frac{d^2\xi}{dt^2} + \nu\frac{d\xi}{dt} + \omega_p^2\xi = F_0 e^{-i\omega t} \tag{4.47}$$

with ξ its displacement from its equilibrium position and ν a damping coefficient. The response, after transient effects have died away, is

$$\xi = \frac{F_0 e^{-i\omega t}}{-\omega^2 - i\nu\omega + \omega_p^2}$$

and the rate at which work is done on the oscillator, averaged over the oscillation period, is

$$\tfrac{1}{2} Re\left(\frac{-i\omega |F_0|^2}{\omega_p^2 - i\nu\omega - \omega^2}\right) = \frac{1}{2} \frac{\nu\omega^2 |F_0|^2}{(\omega_p^2 - \omega^2) + \nu^2\omega^2}$$

If we now suppose that there is a density n_0 of such oscillators, and that their natural frequency varies according to $\omega_p^2 = \omega^2(1 + x/a)$, we find, on integrating over x, that the total absorption per unit area in the y–z plane is

$$\frac{1}{2}\int_{-\infty}^{\infty} \frac{n_0 \nu\omega^2 a^2 |F_0|^2}{\omega^4(x^2 + \nu^2 a^2/\omega^2)} \mathrm{d}x = \frac{\pi n_0 a |F_0|^2}{\omega}$$

which is, indeed, independent of ν.

The absorption per unit volume does depend on ν in the way indicated in Figure 4.13, but the area under the curve is constant. Thus if the damping is small the absorption is highly localized at the critical surface, while if it is somewhat larger there is a broader region of absorption. As the damping tends to zero the solution of (4.47) with $\omega_p = \omega$ has a secular term, so the amplitude increases with time. In this limit there is no solution going as $e^{-i\omega t}$, but instead the problem must be treated as an initial value problem and the energy is fed into an ever-growing wave localized around the critical surface.

As a final comment on resonant absorption we point out that the effect of finite temperature on the plasma wave is to make the group velocity non-zero, as is shown in Chapter 5. As regards our present problem this means that even in the absence of damping energy propagates away from the critical

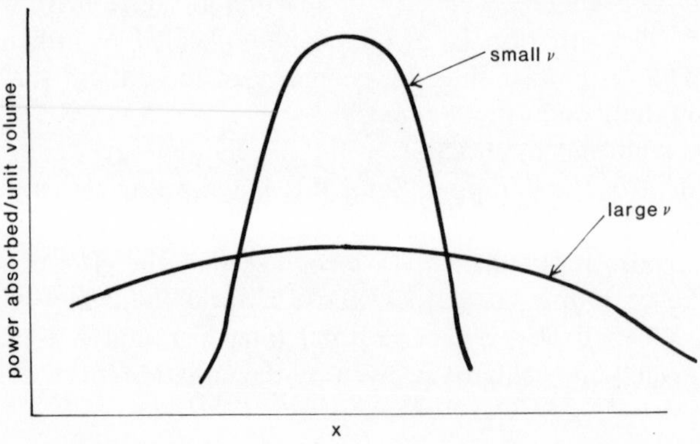

Figure 4.13 Variation of absorption profile with damping.

surface, and the equations no longer have a singularity (but are more complicated). It has, however, been shown that the absorption coefficient of the electromagnetic wave is very little changed.

Problems

4.1 Show that for fields varying as $\exp(i\boldsymbol{k}\cdot\boldsymbol{r}-i\omega t)$ the equations $\nabla\cdot\boldsymbol{E}=\rho/\varepsilon_0$ and $\nabla\cdot\boldsymbol{B}=0$ follow from (4.6) and (4.7). It is not, therefore, necessary to introduce them in the study of plane waves.

4.2 Show that (4.20) reduces to the Alfvén wave dispersion relation at low frequencies, for either sign in the denominators. Hint: Expand in a Taylor series in ω/Ω_i, ω/Ω_e, and keep terms up to second order.

4.3 If for the whistler mode it is assumed that $\Omega_i \ll \omega \ll \Omega_e$ and also that the phase velocity is much less than c, show that its dispersion relation is approximately

$$\omega=\frac{k^2c^2|\Omega_e|}{\omega_{pe}^2}$$

Hence show that its phase and group velocities behave as described in the text.

4.4 Show that near the electron cyclotron frequency the transverse and longitudinal components of the electric field, in a wave propagating perpendicular to a steady magnetic field in the X-mode, are almost equal, while near the upper hybrid frequency the wave is almost entirely longitudinal.

4.5 Show that the energy density in the electromagnetic wave in an unmagnetized plasma is just $\frac{1}{2}\varepsilon_0|E|^2$.

4.6 From (4.16), with only electron terms included, show that

$$n^2=1-\frac{2\alpha\omega^2(1-\alpha)}{2\omega^2(1-\alpha)-\Omega_e^2\sin^2\theta\pm\Omega_e\Gamma}$$

with

$$\Gamma=(\Omega_e^2\sin^4\theta+4\omega^2(1-\alpha)^2\cos^2\theta)^{1/2}$$

and

$$\alpha=\omega_{pe}^2/\omega^2.$$

This is known as the Appleton–Hartree dispersion relation and was first obtained in order to study radio waves propagating in the ionosphere. Hint: Express the roots of the quadratic in n

$$An^4-Bn^2+C=0$$

in the form

$$n^2=1-\frac{2(A-B+C)}{2A-B\pm\sqrt{(B^2-4AC)}}.$$

4.7 If a plasma contains two distinct ion species, show that for propagation across the field, as discussed in the text for the electron cyclotron frequency range, a resonance occurs at a frequency which lies between the two ion cyclotron frequencies. This two-ion hybrid resonance plays an important role in plasma heating schemes in the ion cyclotron frequency range.

4.8 If, in the system considered in section 4.8, the wave has its electric field perpendicular to the plane of incidence show that the equation for the electric field is

$$\frac{d^2E}{dx^2} + k_0^2\left(\frac{x}{a} - \sin^2\theta\right)E = 0.$$

With the transformation

$$y = \left(\frac{k_0^2}{a}\right)^{1/3}(x - a\sin^2\theta)$$

show that this reduces to Airy's equation, demonstrating that there is no resonant absorption for this polarization.

5 Kinetic theory of waves

5.1 Introduction

In this chapter we discuss waves in a hot collisionless plasma, extending the discussion given in Chapter 4 to take account of thermal effects. A proper treatment of the wealth of phenomena occurring cannot be given using fluid equations and we must use a kinetic description. In section 5.2 we shall introduce the basic equation of this work, the Vlasov equation, and show in section 5.3 how when applied to the description of small-amplitude plasma waves it predicts the phenomenon of Landau damping. This damping, the result of a resonant interaction between a wave and particles moving at its phase velocity, is of fundamental importance in plasma physics and is the key to understanding a wide variety of instability mechanisms, as well as mechanisms by which energy can be absorbed from an incident wave. Some of these will be discussed in later sections of the chapter, as well as various extensions of the theory, in particular the inclusion of a magnetic field.

5.2 The Vlasov equation

As discussed briefly in Chapter 1, a proper description of small-scale, high-frequency phenomena in a plasma requires a kinetic description in which we deal with the density of particles in a six-dimensional phase space with three space and three velocity coordinates. Since this is time dependent our description involves a distribution function $f_s(\boldsymbol{r}, \boldsymbol{v}, t)$ of seven variables, s labelling the particle species. The usual fluid variables are moments of this with respect to the velocity; for example the density is

$$n_s(\boldsymbol{r}, t) = \int f(\boldsymbol{r}, \boldsymbol{v}, t)\, \mathrm{d}^3 v \tag{5.1}$$

and the fluid velocity

$$\boldsymbol{v}_s = \frac{1}{n_s} \int \boldsymbol{v} f(\boldsymbol{r}, \boldsymbol{v}, t)\, \mathrm{d}^3 v. \tag{5.2}$$

Sometimes, particularly when a uniform plasma is being considered, the integral (5.1) is taken to have the value 1 so that f_s represents a probability density. However, we shall use the normalization of (5.1).

The Vlasov equation can be obtained as the lowest-order approximation, valid for short time scales, in a systematic treatment of plasma kinetic theory, starting from the equations of motion of a large number of interacting particles and assuming that the number of particles in the Debye sphere is a large parameter. Here, however, we shall take an informal approach, simply seeking to establish this equation as a plausible description of the plasma and to make clear the physics which it contains.

We begin by noting that since particles are conserved the distribution must obey a six-dimensional analogue of the usual fluid equation expressing conservation of mass. Thus,

$$\frac{\partial f_s}{\partial t}+\frac{\partial}{\partial \boldsymbol{r}}\cdot(\dot{\boldsymbol{r}}f_s)+\frac{\partial}{\partial \boldsymbol{v}}\cdot(\dot{\boldsymbol{v}}f_s)=0. \tag{5.3}$$

Here $\partial/\partial \boldsymbol{r}$ denotes the usual gradient operator while $\partial/\partial \boldsymbol{v}$ is the corresponding velocity space operator, $(\partial/\partial v_x, \partial/\partial v_y, \partial/\partial v_z)$ in Cartesian coordinates. The last two terms, taken together, represent the divergence of the particle flux in the six-dimensional phase space. If we now use the equations

$$\dot{\boldsymbol{r}}=\boldsymbol{v} \tag{5.4}$$

$$\dot{\boldsymbol{v}}=\frac{q_s}{m_s}(\boldsymbol{E}+\boldsymbol{v}\times\boldsymbol{B})$$

we obtain from (5.3)

$$\frac{\partial f_s}{\partial t}+\boldsymbol{v}\cdot\frac{\partial}{\partial \boldsymbol{r}}f_s+\frac{q_s}{m_s}(\boldsymbol{E}+\boldsymbol{v}\times\boldsymbol{B})\cdot\frac{\partial}{\partial \boldsymbol{v}}f_s=0, \tag{5.5}$$

which is the Vlasov equation for the distribution function describing the species s. The electric and magnetic fields have to be found from Maxwell's equations and so depend on f through the charge and current, given by

$$\rho=\sum_s q_s\int f_s\,\mathrm{d}^3v \tag{5.6}$$

$$\boldsymbol{J}=\sum_s q_s\int \boldsymbol{v}f_s\,\mathrm{d}^3v.$$

A complete description in our present approximation consists, therefore, of an equation of the form (5.5) for each species with the fields themselves dependent on the f_s through Maxwell's equations and (5.6). The fact that the fields in (5.5) themselves depend on f_s makes the system nonlinear, with the result that it is both difficult to analyse and capable of showing a rich variety of behaviour.

In the above we have effectively replaced the real plasma consisting of discrete particles with a smeared-out continuous density in $(\boldsymbol{r}, \boldsymbol{v})$ space. This might be expected to be reasonable if each particle feels the effect of many other particles simultaneously, and not just that of a few nearest neighbours. As discussed in Chapter 1 this is what happens if there are many particles in the Debye sphere, so that this is the basic condition which must be satisfied if the Vlasov equation is to be a valid description.

It is clear that, in the absence of electric and magnetic fields, an arbitrary function of velocity alone is a solution. On the other hand, we would expect that eventually the velocity distribution would relax to its thermal equilibrium Maxwellian form. In fact this does happen but requires the effect of two-body collisions to be introduced into the equation, for instance through adding the Fokker–Plank collision term obtained in Chapter 3. The Vlasov equation is useful for describing oscillations and other phenomena taking place on time scales much shorter than the collision time associated with relaxation to thermal equilibrium.

5.3 The linearized Vlasov equation and Landau damping

We shall begin our study of the Vlasov equation by looking at what might seem the simplest and most obvious problem, namely that of small-amplitude plasma waves in a uniform plasma with no magnetic field. We shall consider only electron motion, assuming that the ions form an immobile, neutralizing background, and look for electrostatic waves, like the plasma waves discussed in Chapter 4. In these waves there is only an electric field, which can be found from Poisson's equation, and no magnetic field.

Our basic equations are

$$\frac{\partial f_e}{\partial t} + \boldsymbol{v}\cdot\frac{\partial f_e}{\partial \boldsymbol{r}} - \frac{e}{m}\boldsymbol{E}\cdot\frac{\partial f_e}{\partial \boldsymbol{v}} = 0, \tag{5.7}$$

with $\boldsymbol{E}$ found from

$$\boldsymbol{E} = -\nabla\phi, \quad \nabla^2\phi = -\frac{e}{\varepsilon_0}\left(n_i - \int f_e(\boldsymbol{v})\,\mathrm{d}^3v\right). \tag{5.8}$$

As usual in the study of small-amplitude waves we shall take f_e to be a uniform background, with the number density equal to n_i (for singly-charged ions), plus a small perturbation, so that $f_e(\boldsymbol{r}, \boldsymbol{v}, t) = f_0(\boldsymbol{v}) + f_1(\boldsymbol{r}, \boldsymbol{v}, t)$. If we assume that the electric field is zero in the unperturbed state, so that E is a small quantity, we obtain on linearizing the equations

$$\frac{\partial f_1}{\partial t} + \boldsymbol{v}\cdot\frac{\partial f_1}{\partial \boldsymbol{r}} - \frac{e}{m}\boldsymbol{E}\cdot\frac{\partial f_0}{\partial \boldsymbol{v}} = 0 \tag{5.9}$$

$$\nabla^2\phi = \frac{e}{\varepsilon_0}\int f_1\,\mathrm{d}^3v. \tag{5.10}$$

The procedure which we have followed in studying small-amplitude waves with fluid approximations is to assume that all quantities go as $e^{i\boldsymbol{k}\cdot\boldsymbol{r} - i\omega t}$. If we do this with (5.9) and (5.10) then we obtain

$$(-i\omega + i\boldsymbol{k}\cdot\boldsymbol{v})f_1 + \frac{e}{m}i\phi\boldsymbol{k}\cdot\frac{\partial f_0}{\partial \boldsymbol{v}} = 0$$

and

$$-k^2\phi = \frac{e}{\varepsilon_0}\int f_1\,\mathrm{d}^3v.$$

Solving the first of these for f and substituting into the integral in the second we come to the conclusion that if ϕ is non-zero we must have

$$1 + \frac{e^2}{\varepsilon_0 m k^2}\int\frac{\boldsymbol{k}\cdot\partial f_0/\partial\boldsymbol{v}}{\omega - \boldsymbol{k}\cdot\boldsymbol{v}}\,\mathrm{d}^3v = 0. \tag{5.11}$$

If we attempt to take this as the dispersion relation for the waves and use it to find ω in terms of $\boldsymbol{k}$, or vice versa, in the usual way, we come up against the problem that the integral has a singularity on the surface in velocity space where $\omega = \boldsymbol{k}\cdot\boldsymbol{v}$ and is not properly defined.

The way around this problem was first pointed out by Landau in 1946 in a very important paper which laid the basis for much of the subsequent work on plasma oscillations and instabilities. He showed that the problem must be treated as an initial value problem in time in which f_1 is given at $t = 0$ and found at later times, rather than simply assumed to have an $\mathrm{e}^{-i\omega t}$ dependence. We may still Fourier analyse with respect to $\boldsymbol{r}$, so we write

$$f_1(\boldsymbol{r}, \boldsymbol{v}, t) = f_1(\boldsymbol{v}, t)\mathrm{e}^{i\boldsymbol{k}\cdot\boldsymbol{r}}$$

It is useful to take u to be the velocity component along k, i.e. $\boldsymbol{u} = \boldsymbol{k}\cdot\boldsymbol{v}/k$, and define $F_0(u)$, $F_1(u, t)$ to be the integrals of $f_0(\boldsymbol{v})$ and $f_1(\boldsymbol{v}, t)$ over the velocity components perpendicular to $\boldsymbol{k}$. Then we obtain

$$\frac{\partial F_1}{\partial t} + ikuF_1 - \frac{e}{m}E\frac{\partial F_0}{\partial u} = 0 \tag{5.12}$$

and

$$ikE = -\frac{e}{\varepsilon_0}\int_{-\infty}^{\infty} F_1(u)\,\mathrm{d}u \tag{5.13}$$

In order to solve the initial value problem for (5.12) and (5.13) we introduce the Laplace transform of F_1 with respect to t, defined by

$$\tilde{F}_1(u, p) = \int_0^{\infty} F_1(u, t)\,\mathrm{e}^{-pt}\,\mathrm{d}t. \tag{5.14}$$

If the growth of F with t is no faster than exponential, then the integral in (5.14) will converge and define F as an analytic function of p, provided that the real part of p is large enough.

Noting that the Laplace transform of $\mathrm{d}F_1/\mathrm{d}t$ is $p\tilde{F} - F_1(t = 0)$ (as is easily shown by integration by parts), we can Laplace transform (5.12) and (5.13) to obtain

$$p\tilde{F}_1 + iku\tilde{F}_1 = -\frac{e}{m}\tilde{E}\frac{\partial F_0}{\partial u} + F_1(t=0)$$

and

$$ik\tilde{E} = -\frac{e}{\varepsilon_0}\int_{-\infty}^{\infty} \tilde{F}_1(u)\,\mathrm{d}u.$$

From these we have

$$ik\tilde{E} = -\frac{e}{\varepsilon_0}\int_{-\infty}^{\infty}\left\{\frac{-\dfrac{e}{m}\tilde{E}\dfrac{\partial F_0}{\partial u}}{p+iku} + \frac{F_1(t=0)}{p+iku}\right\}\mathrm{d}u,$$

giving

$$\tilde{E} = \frac{1}{ik\varepsilon(k,p)}\int_{-\infty}^{\infty}\frac{F_1(t=0)}{p+iku}\,\mathrm{d}u, \tag{5.15}$$

where

$$\varepsilon(k,p) = 1 - \frac{e^2}{\varepsilon_0 mk}\int_{-\infty}^{\infty}\frac{\dfrac{\partial F_0}{\partial u}}{ip-ku}\,\mathrm{d}u \tag{5.16}$$

The function $\varepsilon(k, p)$ of (5.16) is known as the plasma dielectric constant and, if p is replaced with $-i\omega$, can be seen to be formally the same as the left-hand side of (5.11). However, p now has a positive real part so that the integral is well defined.

The Laplace transform of the distribution function is

$$\begin{aligned}\tilde{F}_1 &= -\frac{e}{m}\frac{\tilde{E}\dfrac{\partial F_0}{\partial u}}{p+iku} + \frac{F_1(t=0)}{p+iku}\\ &= \frac{e\dfrac{\partial F_0}{\partial u}}{ikm\varepsilon(k,p)(p+iku)}\int_{-\infty}^{\infty}\frac{F_1(u',t=0)}{p+iku'}\,\mathrm{d}u' + \frac{F_1(u,t=0)}{p+iku}.\end{aligned} \tag{5.17}$$

Having found the Laplace transforms by means of quite straightforward algebraic manipulations, we must now invert them in order to find the time-dependence of the field and distribution function. The inverse of the Laplace transform is given by

$$F_1(u,t) = \frac{1}{2\pi i}\int_C \tilde{F}_1(u,p)\,\mathrm{e}^{pt}\,\mathrm{d}p, \tag{5.18}$$

where C is a contour parallel to the imaginary axis lying to the right of all singularities of F in the complex p plane, as shown in Figure 5.1.

Rather than try to find a general expression for $F_1(u, t)$ from (5.17) and (5.18) we concentrate on the behaviour at sufficiently large times. Looking at the

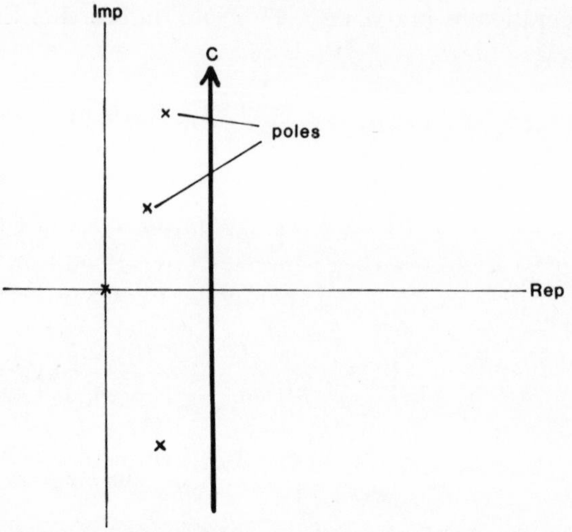

Figure 5.1 Integration contour for inversion of Laplace transforms.

contour of Figure 5.1, we note that if $\tilde{F}(u, p)$ has only a finite number of simple poles in some region $Rep > \sigma$, then we may deform the contour as shown in Figure 5.2, with a loop around each of these singularities. A pole at p_0 gives a contribution going as $e^{p_0 t}$ while the vertical part of the contour, if close to $Rep = -\sigma$, goes as $e^{-\sigma t}$. For sufficiently long times this latter contribution is negligible and the behaviour is dominated by contributions from the poles furthest to the right.

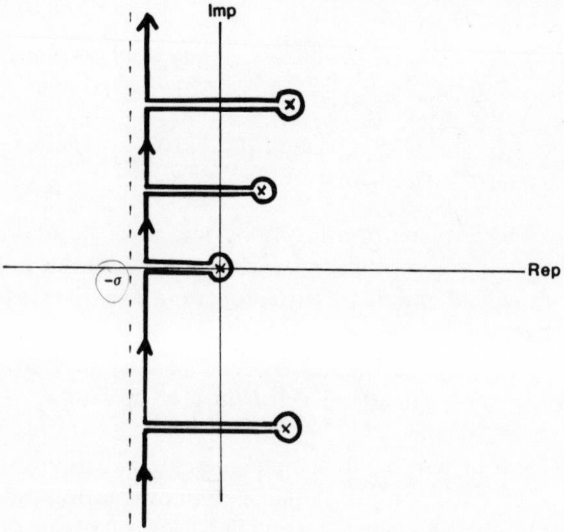

Figure 5.2 Deformed contour for Laplace inversion.

If we examine the expressions (5.15)–(5.17) we see that they all involve integrals of the form

$$\int_{-\infty}^{\infty} \frac{G(u)}{u-(ip/k)}\,du \tag{5.19}$$

which become singular as p approaches the imaginary axis. In order to shift the contour as shown above, we must continue these integrals smoothly across this axis. By virtue of the way in which the Laplace transform is originally defined for $Re\, p$ great enough, the appropriate way to do this is to take the value of these integrals when p is in the right-hand half plane, and find the analytic continuation into the left-hand half plane.

If $G(u)$ is sufficiently well-behaved that it can be continued off the real axis as an analytic function of a complex variable u, then the continuation of (5.19) as the singularity crosses the real axis is obtained by letting it take the contour with it, as shown in Figure 5.3.

The ability to deform the contour into that of Figure 5.2 and find a dominant contribution from a few poles thus depends on F_0 and $F_1(t=0)$ having sufficiently smooth velocity dependences, so that the integrals of (5.15)–(5.17) can be continued sufficiently far into the left-hand half of the complex p plane.

If we consider the electric field given by the inversion of (5.15), we see that its behaviour is dominated by the zero of $\varepsilon(k,p)$ which is furthest to the right, while F_1 has a similar contribution as well as a contribution going as e^{ikut}. Thus, for sufficiently large times after the initiation of the wave, the macroscopic behaviour measured by the electric field associated with the wave depends only on the position of the roots of $\varepsilon(k,p)=0$. The distribution function has a corresponding part, but also a part going as e^{ikut}. For large times this is a rapidly oscillating function of velocity, and its contribution to the charge density, obtained by integrating over u, is negligible.

As we have already noted, the function $\varepsilon(k, p)$ is the same as that of equation (5.11) if p is replaced by $-i\omega$. Thus, the dispersion relation obtained by

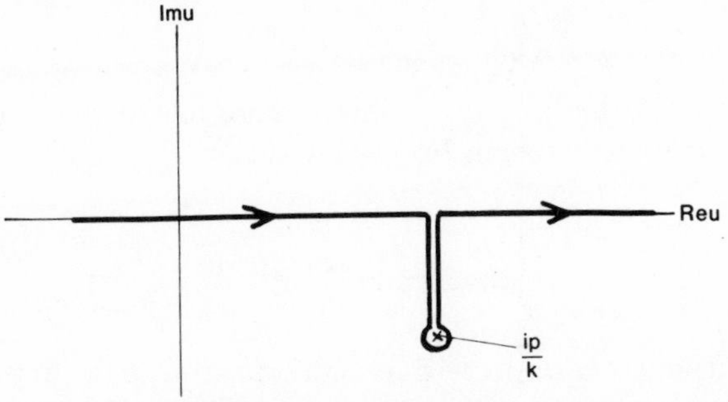

Figure 5.3 Deformation of contour in u plane to give analytic continuation of integrals of the form (5.19).

substituting $e^{i\mathbf{k}\cdot\mathbf{r}-i\omega t}$ will give the correct behaviour for large times, provided that the singular integral is treated correctly. Adapting the procedure which we have found in terms of p we see that the integral must be defined as it stands for $Im\omega > 0$ and analytically continued, by deforming the contour in the u-plane as before, into the region $Im\omega \leqslant 0$. Whether the time-dependence is initially assumed to be as $e^{\pm i\omega t}$, the simplest way to recall how to do the analytic continuation is to note that the integral is continued from the part of the plane corresponding to growing perturbations into that corresponding to damped perturbations. In more complicated problems in which the same difficulty arises, application of this rule allows a much simpler treatment than would be obtained by using Laplace transforms from the beginning.

If we look at the behaviour of the wave dispersion relation in the last chapter we see that for any given k there are a finite number of values of ω, say ω_1, $\omega_2, \ldots$, and any solution is a superposition of functions going as $e^{-i\omega_1 t}$, $e^{-i\omega_2 t}, \ldots$. This set of values of ω is called the spectrum, and the fluid equations give a discrete spectrum. On the other hand in the kinetic problem we obtain contributions to the distribution function going as e^{ikut}, with u taking any real value. The mathematical difficulties of this problem arise from the existence of this continuous spectrum. At short times the behaviour is very complicated and depends on the details of the initial perturbation, and it is only asymptotically that a perturbation going as $e^{-i\omega t}$ is obtained with ω given by a dispersion relation depending only on the unperturbed state. As we have seen, the emergence of such a mode is dependent on the initial velocity perturbation being sufficiently smooth.

Now let us look at the nature of this asymptotic behaviour in the case of a Maxwellian background plasma. If we work in terms of ω rather than p the dispersion relation to be solved is

$$\varepsilon(k, \omega) = 1 + \frac{e^2}{\varepsilon_0 mk} \int_{-\infty}^{\infty} \frac{\dfrac{\partial F_0}{\partial u}}{\omega - ku} \, du = 0 \tag{5.20}$$

with

$$F_0(u) = \frac{n_0}{(2\pi\kappa T/m)^{1/2}} e^{-mu^2/2\kappa T}.$$

Let us suppose that, to a first approximation, ω is real so that, letting ω tend to the real axis from the domain $Im\omega > 0$ we have

$$\int_{-\infty}^{\infty} \frac{\dfrac{\partial F_0}{\partial u}}{\omega - ku} \, du = P \int \frac{\dfrac{\partial F_0}{\partial u}}{\omega - ku} \, du - \frac{i\pi}{k} \left(\frac{\partial F_0}{\partial u} \right)_{u=\omega/k} \tag{5.21}$$

where the P denotes the principal part of the integral (see the Appendix if this is not a familiar concept). Now, if k is taken to be small we may assume that over the range of u where $\partial F_0/\partial u$ is non-negligible $\omega \gg ku$, so that we can expand the denominator of the principal part integral in a Taylor series, i.e.

$$\frac{1}{\omega - ku} \approx \frac{1}{\omega}\left(1 + \frac{ku}{\omega} + \frac{k^2u^2}{\omega^2} + \frac{k^3u^3}{\omega^3} + \cdots\right).$$

Integrating the result term by term and remembering that $\partial F_0/\partial u$ is an odd function, we obtain

$$1 - \frac{\omega_p^2}{\omega^2} - 3k^2\frac{\kappa T\omega_p^2}{m\omega^4} - \frac{e^2}{\varepsilon_0 m}\frac{i\pi}{k^2}\left(\frac{\partial F_0}{\partial u}\right)_{u=\omega/k} \approx 0. \tag{5.22}$$

Equating the real part of (5.22) to zero gives

$$\omega^2 \approx \omega_p^2(1 + 3k^2\lambda_D^2), \tag{5.23}$$

where $\lambda_D = (\kappa T/m\omega_p^2)^{1/2}$ is the Debye length and it is assumed that $k\lambda_D \ll 1$. Then, regarding the imaginary part as producing a small perturbation to this and putting $\omega = \omega_0 + \delta\omega$ with ω_0 the root of (5.23), we have

$$2\omega_0\delta\omega \approx \frac{i\pi\omega_0^2}{k^2}\frac{e^2}{\varepsilon_0 m}\left(\frac{\partial F_0}{\partial u}\right)_{u=\omega/k}$$

and so

$$\begin{aligned}\delta\omega &\approx \frac{i}{2}\frac{\pi}{\omega_p k^2}\frac{e^2}{\varepsilon_0 m}\left(\frac{\partial F_0}{\partial u}\right)_{u=\omega/k}\\ &= -\frac{i}{2}\left(\frac{\pi}{2}\right)^{1/2}\omega_p\frac{1}{k^3\lambda_D^3}\mathrm{e}^{-1/(2k^2\lambda_D^2)-3/2}\end{aligned} \tag{5.24}$$

for Maxwellian F_0.

Comparing the results found here with those for a cold plasma we see first that according to (5.23) ω now depends on k, so that we have a propagating wave with non-zero group velocity. Secondly we have an imaginary part to ω given by (5.24), corresponding, since it is negative, to damping of the wave which is known as Landau damping. If $k\lambda_D \ll 1$, i.e. the wavelength is much larger than the Debye length, then the imaginary part of ω is small compared to the real part and the wave is lightly damped. As, however, the wavelength becomes comparable to the Debye length, the damping becomes strong with the real and imaginary parts comparable. The approximate solution given above is not very accurate in this case, but for many purposes it suffices to note that such waves are very strongly damped.

In the Vlasov equation no dissipative processes are included and it can be verified that if particle velocities are reversed at any time the solution up to that point should simply be reversed in time. At first sight this reversible behaviour does not seem to accord with the fact that an initial perturbation dies out. However, we should note that it is only the electric field which decays and that the distribution function contains undamped terms going as e^{ikut}. Also, the decay of the electric field depends on there being a sufficiently smooth initial perturbation. The presence of the e^{ikut} terms means that as time advances the velocity space dependence of the perturbation becomes more and more convoluted, so that if we reverse the velocities after some time we are not starting with a smooth perturbation. Under these circumstances there is no contradiction in the fact that under time reversal the field will grow initially until the initial state is recreated, and subsequently decay.

Let us now consider briefly the physical reasons for Landau damping, which has appeared only as the result of a rather complicated mathematical analysis. The behaviour of a particle in an electric field varying as $\exp(ikx - i\omega t)$ is given by

$$\frac{d^2x}{dt^2} = \frac{e}{m} E_0 \, e^{ikx - i\omega t}. \tag{5.25}$$

Remembering that we are dealing with a linearized theory in which the perturbation from the wave is small, if the particle starts with velocity u_0 at x_0 we may put $x_0 + u_0 t$ for x in the electric field term. This is the position of the particle on its unperturbed orbit starting at x_0 at time zero. Thus we obtain

$$\frac{du}{dt} = \frac{e}{m} E_0 \, e^{ikx_0 + iku_0 t - i\omega t}$$

and so

$$u - u_0 = \frac{e}{m} E_0 \left\{ \frac{e^{ikx_0 + iku_0 t - i\omega t} - e^{ikx_0}}{iku_0 - i\omega} \right\}. \tag{5.26}$$

As $ku_0 - \omega \to 0$ equation 5.26 tends to

$$u - u_0 = \frac{e}{m} E_0 t \, e^{ikx_0},$$

showing that particles with u_0 close to ω/k, that is with velocity components along x close to the wave phase speed, have velocity perturbations which grow in time. These so-called resonant particles gain energy from or lose energy to the wave and are responsible for the damping, indicating why the damping rate given in (5.24) depends on the slope of the distribution function at $u = \omega/k$. The rest of the particles are non-resonant and have an oscillatory response to the wave field.

To see why energy should be transferred from the electric field to the particles requires more detailed consideration. Whether the speed of a resonant particle increases or decreases depends on the phase of the wave at its initial position and it is not the case that all particles moving slightly faster than the wave lose energy while those moving more slowly than the wave gain energy. Also, the density perturbation is out of phase with the wave electric field, so there is no excess of particles gaining or losing energy initially. If, however, we consider those particles which start with velocities a little above the wave phase velocity, then if they gain energy from the wave they move away from the resonant velocity, while if they lose energy they approach the resonant velocity. The result is that the particles which lose energy interact more effectively with the wave and on average there is a transfer of energy from particles to field in the case of particles starting slightly above the wave speed. The opposite is true for particles with initial velocities just below the wave phase velocity. In the case of a Maxwellian there are more particles in the latter class so there is a net damping. In the limit as the amplitude tends to zero and velocity perturbations due to the wave tend to zero it can be seen that the gradient of the distribution function at the wave speed is what determines the damping rate.

It is of interest to consider the limitations of this result, in terms of the magnitude of the initial field which is allowed before it is seriously in error and nonlinear effects become important. This can be done in terms of quite simple calculations as follows. The basic requirement is that a resonant particle should maintain its position in relation to the phase of the electric field over a time long enough for the damping to take place. To obtain a condition that this be the case consider the problem in a frame of reference in which the wave is at rest and the potential $-e\phi$ seen by an electron is as in Figure 5.4.

If an electron starts at rest, that is in resonance with the wave, at x_0 then it begins to move towards the potential minimum as shown. Validity of the linear theory of Landau damping requires that the position of the particle not change much over the time required for the wave to damp. The time for the particle to change its position with respect to the wave may be estimated as the period required for it to bounce back and forward in the potential well. Near the bottom of the well its equation of motion is approximately

$$\ddot{x} = -\frac{e}{m}k^2 x\phi_0,$$

where k is the wavenumber, and so the bounce time is

$$\tau_b \approx 2\pi\sqrt{\frac{m}{ek^2\phi_0}} = 2\pi\sqrt{\frac{m}{ekE_0}}, \tag{5.27}$$

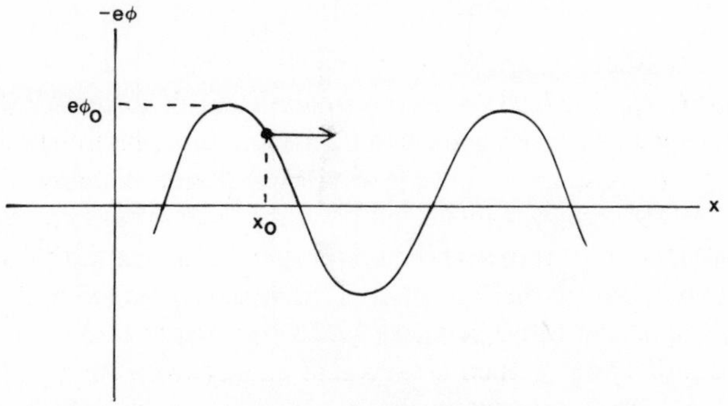

Figure 5.4 Electron potential energy in the rest frame of the wave.

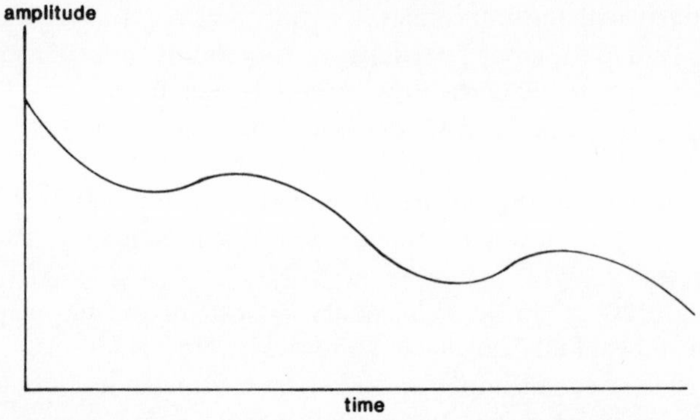

Figure 5.5 Decrease in amplitude of a wave of large initial amplitude.

where E_0 is the field amplitude. We may expect the wave to damp according to the linear theory if τ_b given by (5.27) is much greater than the damping time. Since the former goes inversely as the square root of the electric field amplitude and the latter does not depend on this amplitude this gives an estimate of the maximum initial perturbation which is allowed.

If the initial amplitude is large enough for resonant electrons to be able to bounce back and forth in the potential well a number of times before the wave disappears, then it can be shown that the result to be expected is a non-monotonic decrease in amplitude as shown schematically in Figure 5.5. The period of the amplitude oscillations is around the value τ_b of equation 5.27.

5.4 The plasma dispersion function

If the distribution F_0 in (5.20) is Maxwellian then it is readily seen that with a suitable scaling of the variables the dispersion relation can be expressed in terms of the function

$$Z(\zeta) = \pi^{-1/2} \int_{-\infty}^{\infty} \frac{dt\, e^{-t^2}}{t-\zeta} \tag{5.28}$$

defined in this way for $Im\zeta > 0$ and by analytic continuation for other values. This function is known as the plasma dispersion function and often occurs in problems involving small-amplitude waves in a plasma. The factor $\pi^{-1/2}$ is just a conventional normalization.

In view of its importance and regular occurrence in the literature of plasma physics we shall devote this section to a discussion of its main properties. Let us first note that if we differentiate (5.28) with respect to ζ we get

$$Z'(\zeta) = \pi^{-1/2} \int_{-\infty}^{\infty} \frac{dt\, e^{-t^2}}{(t-\zeta)^2},$$

which yields, on integration by parts

$$Z'(\zeta) = -\pi^{-1/2}\int_{-\infty}^{\infty} \frac{2t}{t-\zeta} e^{-t^2}\, dt$$

$$= -2\pi^{-1/2}\int_{-\infty}^{\infty}\left(1 + \frac{\zeta}{t-\zeta}\right) e^{-t^2}\, dt$$

$$= -2[1 + \zeta Z]. \tag{5.29}$$

If we let ζ tend to zero from the upper half of the complex plane we obtain

$$Z(0) = \pi^{-1/2} P \int_{-\infty}^{\infty} \frac{dt\, e^{-t^2}}{t} + i\pi^{1/2}$$

$$= i\pi^{1/2},$$

the principal part integral being zero because its integrand is an odd function of t.

Integrating the linear differential equation 5.29, which has an integrating factor e^{ζ^2}, and using the above boundary condition, we derive an alternative expression for Z, namely

$$Z(\zeta) = e^{-\zeta^2}\left(i\pi^{1/2} - 2\int_0^{\zeta} e^{x^2}\, dx\right), \tag{5.30}$$

which, by making the substitution $t = ix$ in the integral and noting that

$$\int_{-\infty}^{0} e^{-t^2}\, dt = \tfrac{1}{2}\pi^{1/2}$$

can be written in the form

$$Z(\zeta) = 2i\, e^{-\zeta^2}\int_{-\infty}^{i\zeta} e^{-t^2}\, dt. \tag{5.31}$$

This form, which relates the plasma dispersion function to an error function with argument $i\zeta$, is valid for all values of ζ.

For small ζ we have the expansion

$$Z(\zeta) = i\pi^{1/2} e^{-\zeta^2} - 2\zeta\left[1 - \frac{2\zeta^2}{3} + \frac{4\zeta^4}{15} - \frac{8\zeta^6}{105}\cdots\right]$$

while for large $\zeta = x + iy$ we have for $x > 0$ the following asymptotic expansion

$$Z(\zeta) \sim i\pi^{1/2}\sigma e^{-\zeta^2} - \zeta^{-1}\left[1 + \frac{1}{2\zeta^2} + \frac{3}{4\zeta^4} + \frac{15}{8\zeta^6} + \cdots\right],$$

with

$$\sigma = \begin{cases} 0 & y > 1/x \\ 1 & |y| < 1/x \\ 2 & -y > 1/x \end{cases}$$

In deriving our expression for Landau damping in the last section we have, in effect, used the first few terms of this asymptotic expansion.

5.5 Ion sound waves

If we include ion dynamics in the theory then the dispersion relation (5.20) becomes

$$1+\frac{e^2}{\varepsilon_0 m_e k}\int\frac{\dfrac{\partial F_{0e}}{\partial u}}{\omega-ku}\,du+\frac{e^2}{\varepsilon_0 m_i k}\int\frac{\dfrac{\partial F_{0i}}{\partial u}}{\omega-ku}\,du=0,$$

that is we simply add an extra term of the same form as the electron term. Assuming that we can find a wave with ω/k much less than the electron thermal velocity and much greater than the ion thermal velocity, conditions which we shall check later, we may assume $ku \ll \omega$ in the ion term as we did previously for the electron term, and get this term to be approximately ω_{pi}^2/ω^2. In the electron term we have $ku \gg \omega$ for the values of u of interest, so we neglect ω in the integral and, assuming F_{0e} to be Maxwellian with temperature T_e, obtain

$$\frac{\omega_{pe}^2}{k^2}\frac{m_e}{T_e}=\frac{1}{k^2\lambda_D^2}$$

for this term.

Thus a first approximation to the dispersion relation is

$$1+\frac{1}{k^2\lambda_D^2}+\frac{\omega_{pi}^2}{\omega^2}=0$$

giving

$$\omega^2=\frac{\omega_{pi}^2 k^2\lambda_D^2}{1+k^2\lambda_D^2}=\frac{T_e}{m_i}\frac{k^2}{1+k^2\lambda_D^2}. \tag{5.32}$$

For $k \ll \lambda_D^{-1}$ we have $\omega=(T_e/m_i)^{1/2}k$, an equation which is like that of an ordinary sound wave with the pressure provided by the hot electrons and the inertia by the ions. As the wavelength is reduced towards the Debye length the frequency levels off and approaches the ion plasma frequency.

Checking that our original assumptions are true we see that the wave phase velocity $(T_e/m_i)^{1/2}$ is indeed less than the electron thermal velocity (by a factor $(m_e/m_i)^{1/2}$), but that it is only much greater than the ion thermal velocity if the ion temperature T_i is much less than T_e. If $T_i \ll T_e$ the wave phase velocity can lie on almost flat portions of the ion and electron velocity distributions as shown in Figure 5.6 and be very little Landau damped. The condition that an ion sound wave should propagate without being strongly damped in a distance of the order of its wavelength is usually given as being that T_e should be at least five to ten times T_i.

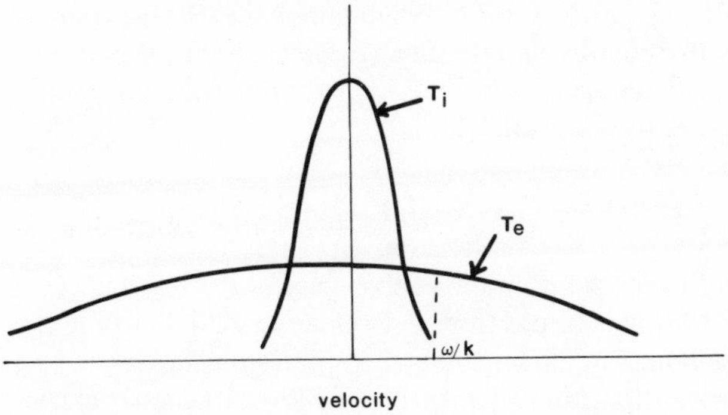

Figure 5.6 Ion sound velocity in relation to particle distribution functions for $T_e \gg T_i$.

5.6 Waves in a magnetized plasma

If we now consider a plasma in a uniform background magnetic field $\boldsymbol{B}_0$ and also take into account magnetic field perturbations so that we no longer look only for electrostatic waves, then the linearized Vlasov equation takes, for either electrons or ions, the form

$$\frac{\partial f_1}{\partial t} + \boldsymbol{v}\cdot\frac{\partial f_1}{\partial \boldsymbol{r}} + \frac{q}{m}(\boldsymbol{v} \times \boldsymbol{B}_0)\cdot\frac{\partial f_1}{\partial \boldsymbol{v}} = -\frac{q}{m}(\boldsymbol{E}_1 + \boldsymbol{v} \times \boldsymbol{B}_1)\cdot\frac{\partial f_0}{\partial \boldsymbol{v}} \tag{5.33}$$

where the field perturbations $\boldsymbol{E}_1$ and $\boldsymbol{B}_1$ are related through Maxwell's equations with the current given by an integral over the distribution function.

Before solving this we should note first that the function f can no longer be an arbitrary function of $\boldsymbol{v}$, but must obey

$$(\boldsymbol{v} \times \boldsymbol{B}_0)\cdot\frac{\partial f_0}{\partial \boldsymbol{v}} = 0$$

a relation which, if we express $\boldsymbol{v}$ in cylindrical polar coordinates $(v_z, v_\perp, \theta)$ can readily be shown to imply that $\partial f_0/\partial\theta = 0$, i.e. f_0 is a function only of v_z and $v_\perp$. To solve (5.33) we observe if the trajectory of a particle in phase space is $\boldsymbol{r} = \boldsymbol{r}(t)$, $\boldsymbol{v} = \boldsymbol{v}(t)$, then in the unperturbed field

$$\frac{\mathrm{d}\boldsymbol{r}}{\mathrm{d}t} = \boldsymbol{v} \quad \text{and} \quad \frac{\mathrm{d}\boldsymbol{v}}{\mathrm{d}t} = \frac{q}{m}(\boldsymbol{v} \times \boldsymbol{B}_0)$$

so that equation (5.33) can be written as

$$\frac{\mathrm{D}f_1}{\mathrm{D}t} = -\frac{q}{m}(\boldsymbol{E}_1 + \boldsymbol{v} \times \boldsymbol{B}_1)\cdot\frac{\partial f_0}{\partial \boldsymbol{v}} \tag{5.34}$$

where $\mathrm{D}f_1/\mathrm{D}t$ is the total rate of change of f_1, following the unperturbed trajectory of the particle. Readers familiar with the theory of partial differential equations will recognize that the unperturbed particle orbits are the characteristics of the differential equation 5.33. Under the assumption that f_1 vanishes as $t \to -\infty$ the solution of (5.34) is

$$f_1(\boldsymbol{r}, \boldsymbol{v}, t) = -\frac{q}{m}\int_{-\infty}^{t} \{\boldsymbol{E}_1(\boldsymbol{r}', t') + \boldsymbol{v}' \times \boldsymbol{B}_1(\boldsymbol{r}', t')\} \cdot \frac{\partial f_0(\boldsymbol{v}')}{\partial \boldsymbol{v}'}, \tag{5.35}$$

where $\boldsymbol{r}'$ and $\boldsymbol{v}'$ are the position and velocity of the particle at time t' on the unperturbed orbit with position and velocity $\boldsymbol{r}$ and $\boldsymbol{v}$ at time t.

It may be noted, in passing, that this solution technique will work for any equilibrium solution, not just one with a uniform magnetic field. However, in a uniform magnetic field we know that the unperturbed orbit is a helix, while in a general field configuration it is difficult to find a closed form for the orbit sufficiently simple to allow further progress.

Using the solution (2.3) of the orbit equations in a uniform magnetic field and expressing $\boldsymbol{v}$ in cylindrical polar coordinates so that $\boldsymbol{v} = (v_\perp \cos\theta, v_\perp \sin\theta, v_z)$ we have

$$\boldsymbol{v}' = (v_\perp \cos(\Omega(t - t') + \theta), v_\perp \sin(\Omega(t - t') + \theta), v_z), \tag{5.36}$$

and on integrating this

$$x' = x - \frac{v_\perp}{\Omega}[\sin(\Omega(t - t') + \theta) - \sin\theta]$$

$$y' = y - \frac{v_\perp}{\Omega}[\cos(\Omega(t - t') + \theta) - \cos\theta]$$

$$z' = z + v_z(t - t'). \tag{5.37}$$

Here we have used the fact that $v_\perp$ and v_z are constants of the motion.

We shall miss out much of the algebraic detail of the subsequent development, since this is readily available from more advanced texts, and concentrate on explaining the essential features of the calculation. The first thing to note is that, since f_0 depends only on the constants of motion $v_\perp$ and v_z, $f_0(\boldsymbol{v}') = f_0(\boldsymbol{v})$ and

$$\frac{\partial f_0}{\partial v'_x} = \cos(\Omega(t - t') + \theta)\frac{\partial f_0}{\partial v_\perp}$$

$$\frac{\partial f_0}{\partial v'_y} = \sin(\Omega(t - t') + \theta)\frac{\partial f_0}{\partial v_\perp}$$

$$\frac{\partial f_0}{\partial v'_z} = \frac{\partial f_0}{\partial v_z}.$$

The next step is to assume that all the perturbation go as $e^{i\boldsymbol{k}\cdot\boldsymbol{r} - i\omega t}$ or, taking the perpendicular component of the wavenumber to be along the x-axis,

$e^{ik_\perp x + ik_z z - i\omega t}$. The terms in the integral in (5.35) going as $e^{ik\cdot r'}$ give terms of the type

$$\exp\left(\frac{ik_\perp v_\perp}{\Omega}\sin(\Omega(t-t')+\theta)\right)$$

which are expanded using the identity

$$e^{ia\sin x} = \sum_{n=-\infty}^{\infty} J_n(a)\, e^{inx}. \tag{5.38}$$

This reduces the integration to a sum of elementary integrals over exponential functions and allows us to evaluate f_1, which is used in turn to find the current density by an integration over velocity. The magnetic field can be eliminated by means of Maxwell's equations, and following the same sort of procedure as in the previous chapter we can calculate the dielectric tensor, the end result being that

$$\varepsilon_{ij} = \delta_{ij} + \sum_s \frac{\varepsilon_0 q_s^2}{m_s \omega^2} \sum_{n=-\infty}^{\infty} \int d^3v \frac{S_{ij}}{\omega - k_z v_z - n\Omega_s}, \tag{5.39}$$

the first sum being over the species present in the plasma and the second coming from (5.38). The elements of the tensor S_{ij} which appears in (5.39) are given by

$$S_{ij} = \begin{pmatrix} v_\perp U\left(\dfrac{nJ_n}{a_s}\right)^2 & -iv_\perp U\dfrac{n}{\lambda_s}J_n J_n' & v_\perp W\dfrac{n}{a_s}J_n^2 \\ iv_\perp U\dfrac{n}{a_s}J_n J_n' & v_\perp U J_n'^2 & iv_\perp W J_n J_n' \\ v_z U\dfrac{n}{a_s}J_n^2 & -iv_z U J_n J_n' & v_z W J_n^2 \end{pmatrix}, \tag{5.40}$$

where

$$U = (\omega - k_z v_z)\frac{\partial f_{0s}}{\partial v_\perp} + k_z v_\perp \frac{\partial f_{0s}}{\partial v_z}$$

$$W = \frac{n\Omega_s v_z}{v_\perp}\frac{\partial f_{0s}}{\partial v_\perp} + (\omega - n\Omega_s)\frac{\partial f_{0s}}{\partial v_z}$$

$$a_s = \frac{k_\perp v_\perp}{\Omega_s}.$$

The argument of the Bessel functions in (5.40) is a_s.

This dielectric tensor is used to investigate the properties of waves in just the same way as that for the cold plasma was used in Chapter 4, although its properties are much more complicated. It will be noted from (5.39) that it involves singular integrals of a type similar to those encountered in section 5.3. In principle the problem ought to be treated as an initial value problem, but we can use our knowledge of the simpler case to recognize that the appropriate way to treat the singular integrals is to evaluate them as written for $Im\omega > 0$ and by analytic continuation for $Im\omega \leqslant 0$.

For Maxwellian distribution functions the integral over $\mathbf{v}$ in (5.39) can be carried out, making use of the identity

$$\int_0^\infty xJ_n^2(sx)\,\mathrm{e}^{-x^2} = \tfrac{1}{2}\mathrm{e}^{-s^2/2}I_n\left(\frac{s^2}{2}\right),$$

where I_n is the modified Bessel function, the result being that

$$\varepsilon_{ij} = \delta_{ij} + \sum_s \frac{\omega_{ps}^2}{\omega^2}\left(\frac{m_s}{2\kappa T_s}\right)^{1/2}\frac{1}{k_z}\mathrm{e}^{-\lambda_s}\sum_{n=-\infty}^{\infty} T_{ij}. \tag{5.41}$$

with

$$T_{ij} = \begin{pmatrix} \dfrac{n^2 I_n Z}{\lambda_s} & -in(I_n' - I_n)Z & \dfrac{nI_n Z'}{(2\lambda_s)^{1/2}} \\ in(I_n' - I_n)Z & \left(\dfrac{n^2 I_n}{\lambda_s} + 2\lambda_s I_n - 2\lambda I_n'\right)Z & \dfrac{\lambda_s^{1/2}(I_n' - I_n)}{2^{1/2}}Z' \\ \dfrac{nI_n Z'}{(2\lambda_s)^{1/2}} & \dfrac{\lambda_s^{1/2}(I_n' - I_n)Z'}{2^{1/2}} & I_n Z \end{pmatrix}. \tag{5.42}$$

In (5.42), λ_s, which is the argument of the Bessel functions and their derivatives, is defined by

$$\lambda_s = \frac{\kappa T_s k_\perp^2}{m_s \Omega_s^2},$$

while Z and Z' are the plasma dispersion function and its derivative with argument

$$\xi_n = \frac{\omega - n\Omega}{k_z}\left(\frac{m_s}{2\kappa T_s}\right)^{1/2}.$$

As $k_z \to 0$, $\xi_n \to \infty$ and $Z(\xi_n) \sim 1/\xi_n$, from which we can see that for perpendicular propagation the dielectric tensor has the same block structure as that for a cold plasma, since the terms involving Z' vanish. Thus we have the same separation into ordinary and extraordinary modes, though in this case the ordinary mode, with its electric field along the z-axis, is not independent of the magnetic field. Some of the main properties of waves in a hot plasma in a magnetic field will be elaborated on in the next two sections.

5.7 Bernstein modes

We shall consider high-frequency modes, allowing us to simplify the dispersion relation by neglecting the ion terms. The subscript s will be dropped, it being understood that all quantities relate to electrons. Letting k_z go to zero, as described above, we obtain, for those waves with non-zero E_x and E_y, the dispersion relation

$$\left(1-\frac{\omega_p^2}{\omega^2}\frac{e^{-\lambda}}{\lambda}\sum_{n=-\infty}^{\infty}\frac{n^2 I_n}{\omega-n\Omega}\right)\left(1-\frac{c^2k^2}{\omega^2}-\frac{\omega_p^2 e^{-\lambda}}{\omega}\sum_{n=-\infty}^{\infty}\frac{n^2I_n/\lambda+2\lambda I_n-2\lambda I_n'}{\omega-n\Omega}\right)$$
$$=\left(\frac{\omega_p^2 e^{-\lambda}}{\omega}\sum_{n=-\infty}^{\infty}\frac{n(I_n'-I_n)}{\omega-n\Omega}\right)^2. \tag{5.43}$$

As $\lambda \to 0$ we recover the cold plasma extraordinary mode dispersion relation. This mode, for which $\lambda \ll 1$ unless the plasma has a thermal velocity approaching the velocity of light, is little affected by thermal effects, except near the cyclotron harmonics $\omega = n\Omega$ where small thermal corrections are important because of the smallness of the denominators in (5.43).

However, another mode also exists. If we look for a mode with phase velocity much less than c, i.e. $c^2k^2/\omega^2 \gg 1$, then from (5.41) we see that its dispersion relation must be approximately

$$1-\frac{\omega_p^2}{\omega}\frac{e^{-\lambda}}{\lambda}\sum_{n=-\infty}^{\infty}\frac{n^2 I_n}{\omega-n\Omega}=0 \tag{5.44}$$

obtained by putting the coefficient of c^2k^2/ω^2 equal to zero. This corresponds to a wave with $E_y = 0$ but non-zero E_x, that is a longitudinal wave. Another way of obtaining its dispersion relation is to note that a longitudinal wave is purely electrostatic and can be obtained by neglecting the displacement current in Maxwell's equations, a process equivalent to letting $c \to \infty$ in the dispersion relation. As $\lambda \to 0$ in (5.44) with $\omega \neq n\Omega$ for any n, then only the $n = \pm 1$ terms contribute. Since $I \pm 1(\lambda)/\lambda \to 1/2$ as $\lambda \to 0$

$$\omega^2 \to \omega_p^2 + \Omega^2,$$

so that there is a solution with $k \to 0$ at the upper hybrid frequency. For other non-zero values of n, $I_n(\lambda)/\lambda \to 0$, but a solution can be obtained with $\omega \to n\Omega$ at the same time. Similarly as $\lambda \to \infty$, $e^{-\lambda}I_n(\lambda) \to 0$, and so to satisfy the equation ω must tend to $n\Omega$ for some n. The complete solution of (5.44) is as shown in Figure 5.7 which is drawn for the upper hybrid frequency between $2|\Omega|$ and $3|\Omega|$. Wherever the upper hybrid frequency lies, the modes above and below it behave like those in this diagram.

These longitudinal modes are usually known as Bernstein modes, after I.B. Bernstein who first showed that they propagated with real values of ω and investigated their properties.

For small k the phase velocity becomes large and it is no longer legitimate to neglect the electromagnetic terms. A more detailed examination of the complete dispersion relation shows that the electromagnetic mode and the longitudinal mode cross over near the harmonics to give the pattern shown in Figure 5.8.

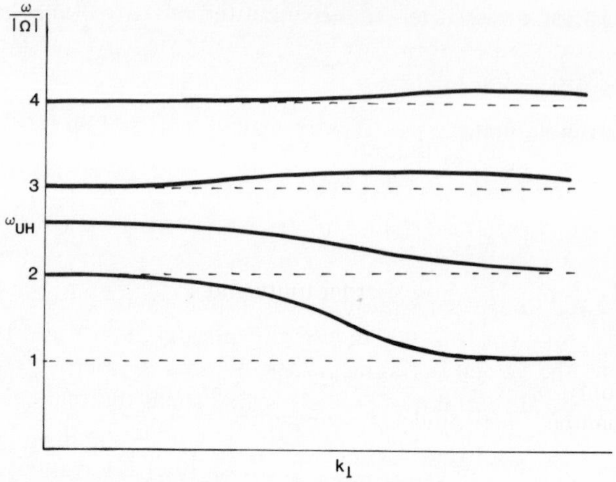

Figure 5.7 Dispersion curves for the Bernstein modes.

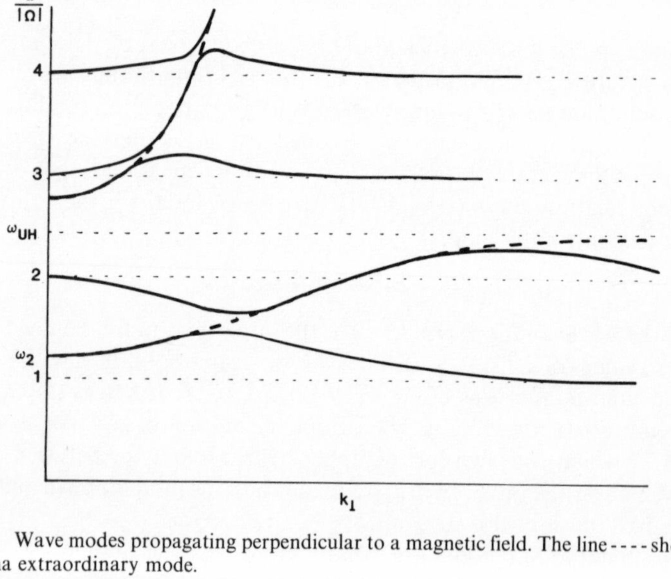

Figure 5.8 Wave modes propagating perpendicular to a magnetic field. The line - - - - shows the cold plasma extraordinary mode.

In a lower frequency range a similar phenomenon occurs at harmonics of the ion cyclotron frequency, producing ion Bernstein modes, with somewhat similar properties. However it must be borne in mind that while the ion contribution to the dispersion relation can be neglected for high-frequency waves, the electron contribution cannot be neglected for low frequencies, so that there is not a complete symmetry between the two types of Bernstein mode.

5.8 Cyclotron damping

For finite k_z the plasma dispersion functions in (5.42) have a non-zero imaginary part with the result that in general ω is complex, giving damping of the waves in a thermal equilibrium plasma. The imaginary part of $Z(\xi_n)$ is very small except when ξ_n is of order unity and so the electromagnetic waves, for which k_z is of order ω/c, are only strongly damped for frequencies in a narrow band around the cyclotron harmonics. The Bernstein mode, which for small k_z only propagates close to the cyclotron harmonics, is generally strongly damped if its angle of propagation is not within a few degrees of perpendicular to the magnetic field.

In Chapter 4 we described some of the radio-frequency heating schemes which are currently of interest in fusion research, and which often involve eventual absorption of a wave by cyclotron damping of the type discussed here. The origin of the damping can be traced back to the resonant denominator $\omega - k_z v_z - n\Omega_s$ in (5.37). The guiding centre of a particle moves along the field with velocity v_z, so in a frame of reference moving with the guiding centre, the wave frequency is Doppler shifted to $\omega - k_z v_z$. There appears then to be a resonant exchange of energy between the particle and the wave when this Doppler shifted frequency is equal to an integer multiple of the cyclotron frequency. To see why all the harmonics of the cyclotron frequency should be important we consider the equation of motion of a particle in a wave whose associated fields vary as $\exp(ik_\perp x + ik_z z - i\omega t)$. This is

$$\frac{d\boldsymbol{v}}{dt} - \frac{q}{m}(\boldsymbol{v} \times \boldsymbol{B}_0) = \frac{q}{m}(\boldsymbol{E}_1 + \boldsymbol{v} \times \boldsymbol{B}_1)\exp(ik_\perp x + ik_z z - i\omega t),$$

and if we regard the right-hand side as a small perturbing term in which we may put the unperturbed values of x and z, the argument of the exponential becomes

$$\frac{ik_\perp v_\perp}{\Omega}(\sin(\Omega_t + \theta) - \sin\theta) + ik_\perp x_0 + ik_z z + ik_z v_z t - i\omega t.$$

With the aid of the identity (5.38) it can be seen that the time dependence of this has Fourier components at frequencies $\omega - k_z v_z - n\Omega$, for all integers n. The particle responds resonantly to a driving force at the cyclotron frequency. If, however, we move with the particle we do not see the field due to the wave as being a simple sinusoidal oscillation at the Doppler shifted frequency $\omega - kv_z$. The oscillations of the particle in the perpendicular direction mean that its position is changing with respect to the phase of the wave and the result is that it sees a field which contains a superposition of frequency components at

$\omega - k_z v_z - n\Omega$. One of these will be in resonance with the particle's cyclotron motion if $\omega - k_z v_z$ is an integral multiple of Ω. The Bessel functions $J_n(k_\perp v_\perp/\Omega)$ which are characteristic of the dispersion relation in a magnetic field are the magnitudes of the various Fourier components of the wave field as seen by an observer moving with the particle. For electromagnetic waves $k_\perp v_\perp/\Omega$ is small compared with one for most particles, in which case, since the Bessel functions go approximately as $(k_\perp v_\perp/\Omega)^n/(2^n n!)$, the cyclotron damping is only important for low harmonics. If $k_z = 0$ the resonance condition is independent of particle velocity, so that all particles of the plasma are resonantly driven at once and the result is that $k_\perp \to \infty$, just as for cold plasma resonances.

With regard to this last point it is interesting to note that for electrons this behaviour is modified by relativistic effects, which make the cyclotron frequency, which depends on the mass, velocity-dependent. The result is that, even at exactly perpendicular incidence, the resonance condition is velocity-dependent and we are back to the Landau damping type of behaviour where only a special group of particles interact strongly with the wave. We shall not pursue this matter here, but it should be borne in mind that relativistic effects are important around electron cyclotron harmonics even in comparatively low-temperature plasmas in which relativistic corrections to electron dynamics are otherwise negligible. This is particularly the case for waves propagating almost perpendicular to the magnetic field.

5.9 Microinstabilities

So far we have discussed the behaviour of waves only in thermal equilibrium plasmas with Maxwellian distributions. Typically Landau or cyclotron damping occurs because of the transfer of energy from waves to a group of particles whose velocities satisfy some resonance condition. The rate of damping depends on the distribution of particles in this special region of velocity space. If the distribution is not Maxwellian there is the possibility that the energy transfer may be modified and go the other way, so that the wave grows at the expense of the particle kinetic energy. This type of instability, which depends on the microscopic details of the velocity distribution function, is generally known as a microinstability, to distinguish it from the type of macroscopic fluid instability discussed in Chapter 3. Microinstabilities generally involve the growth of the kind of small-scale wave looked at in this chapter, so that their effect is not to produce a bulk motion of the plasma, but rather an enhanced level of fluctuations. While not having the drastic effect on confinement of a large-scale fluid instability, they may degrade confinement by producing diffusion rates in excess of those expected in a completely stable plasma.

The simplest example of such an instability is obtained by modifying slightly the Maxwellian distribution which gave the damping given by equation 5.26. If

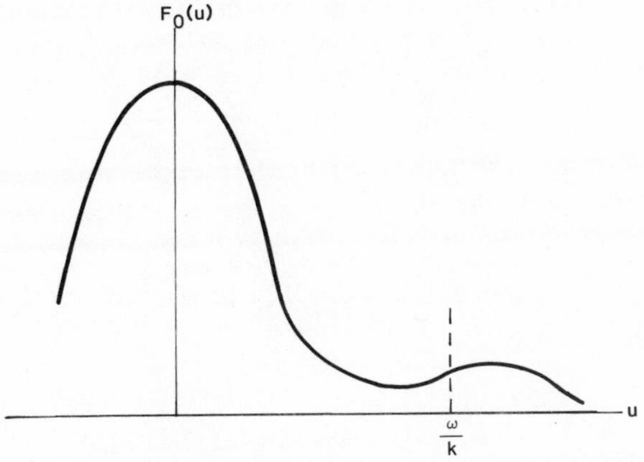

Figure 5.9 Distribution function for the 'bump on tail' instability.

the distribution is modified in the way shown in Figure 5.9, then the real part of ω is not significantly affected by the small change on the tail of the Maxwellian, but the imaginary part, depending on the gradient of $F_0(u)$ at the phase velocity, is changed in sign, giving growth rather than damping of the wave. This 'bump on tail' instability has often been used as a starting point of non-linear theories, aimed at analysing the process which must eventually halt the exponential growth of the perturbation.

For high-frequency electrostatic oscillations, described by a dispersion relation of the form (5.20) with a general distribution function F_0, there is a simple general criterion for instability obtained by O. Penrose. The dispersion relation (5.20) can be written as

$$k^2 = \frac{e^2}{\varepsilon_0 m} \int \frac{\dfrac{\partial F_0}{\partial u}}{u - \dfrac{\omega}{k}} du$$

$$= \frac{e^2}{\varepsilon_0 m} \int \frac{\dfrac{\partial F_0}{\partial u}}{u - U} du$$

$$= f(U) \text{ say.}$$

Taking k to be real and positive, the question of whether the system is unstable or not is equivalent to asking whether this equation is satisfied by any values of U in the upper half of the complex plane. To answer this we use a standard result of complex variable theory which says that the number of zeroes minus the number of poles of $f(U) - k^2$ in a given region of the complex U plane is $(2\pi)^{-1}$ times the increase in the argument of $f(U) - k^2$ when U moves once anticlockwise around the boundary of this region. To find this latter quantity

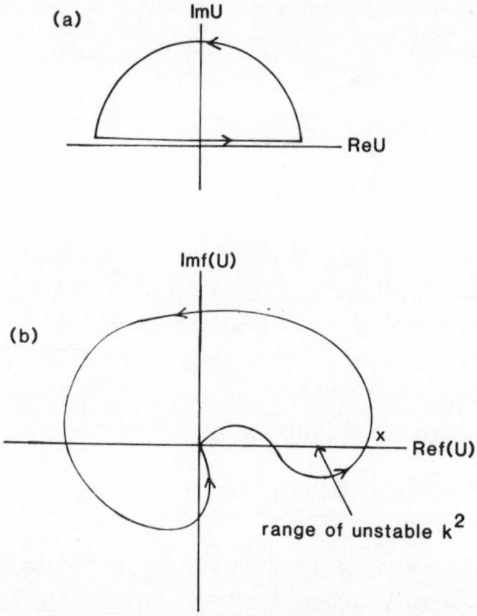

Figure 5.10 The Nyquist diagram. As U moves around the contour in (*a*), $f(U)$ follows that of (*b*).

we construct what is known as a Nyquist diagram. Since the region in which we are interested is the upper half complex plane, we let U follow the path shown in Figure 5.10(*a*) and plot the corresponding path followed in the complex plane by $f(U)$, as illustrated in Figure 5.10(*b*). As $|U| \to \infty f(U) \to 0$, and so as the radius of the semicircle in Figure 5.10(*a*) tends to infinity, only the part of the contour along the real axis is important, and the $f(U)$ contour starts and finishes at the origin. Since $f(U)$ is analytic in the upper half U plane, by virtue of the way in which it is defined, the number of zeroes of $f(U) - k^2$ is the change in argument of this quantity as the path in Figure 5.10(*b*) is followed. This is just the number of times the path encircles the point k^2, so the criterion for instability is that the path encircles part of the positive real axis. If it turned out to be as illustrated in Figure 5.10(*b*), for instance, the system would be unstable for the values of k^2 shown. In an unstable system there must exist a point such as x in Figure 5.10(*b*), where the contour crosses the real axis going from negative to positive real part. As U moves along the real axis

$$\frac{\varepsilon_0 m}{e^2} f(U) = P \int_{-\infty}^{\infty} \frac{F_0'(u)}{u - U} \mathrm{d}u + i\pi F_0'(u)$$

and so at a point such as x corresponding to value $U = U_0$ say, $F_0'(U_0) = 0$ and as U increases through U_0, F_0' goes from negative to positive. This implies that $F_0(U)$ has a minimum at the value U. A further condition to be satisfied is that the real part of $f(U)$ be positive at this point, i.e.

$$\int_{-\infty}^{\infty} \frac{F_0'}{u - U_0} \mathrm{d}u > 0$$

where the principal part need not be taken since the numerator vanishes at the same time as the denominator. As integration by parts yields the equivalent condition

$$\int_{-\infty}^{\infty} \frac{F_0(u) - F_0(U_0)}{(u - U_0)^2} > 0 \tag{5.45}$$

$F_0(U_0)$ having been chosen as a constant of integration in order to make it again unnecessary to take the principal part. The condition that $F_0(u)$ has a minimum implies that $F(u)$ has two or more maxima, while (5.45) can be seen to imply that the minimum must be of sufficient depth. Although the idea of the Nyquist diagram may be applied to any stability problem, for more complicated dispersion relations it is not so easy to obtain a simple criterion to be applied to the distribution function.

An important type of instability occurs at lower frequencies when the electrons and ions drift relative to each other, as would be the case in a plasma carrying a current. If the drift velocity of the electrons is small compared to their thermal velocity, the analysis leading to equation 5.34 for the frequency of the ion sound wave goes through as before, but Figure 5.6 is modified to look like Figure 5.11. Under the same condition as before, that the electron temperature be greater than the ion temperature, the electrons may provide a positive contribution to the growth rate outweighing the negative contribution of the ions. This ion sound instability can occur when the wave phase velocity is less than the drift velocity of the electrons, so that the slope of the electron distribution function is positive at the wave speed. If the electron drift velocity is much greater than the thermal velocity, we obtain the sort of distribution functions shown in Figure 5.12.

The dispersion relation for electrostatic waves in this system is approximately

$$\frac{\omega_{pi}^2}{\omega^2} + \frac{\omega_{pe}^2}{(\omega - \boldsymbol{k}\cdot\boldsymbol{U})^2} = 1 \tag{5.46}$$

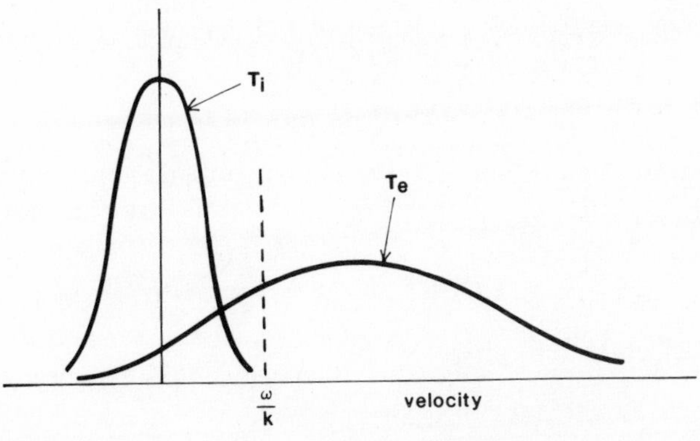

Figure 5.11 Shifted electron distribution producing ion sound instability.

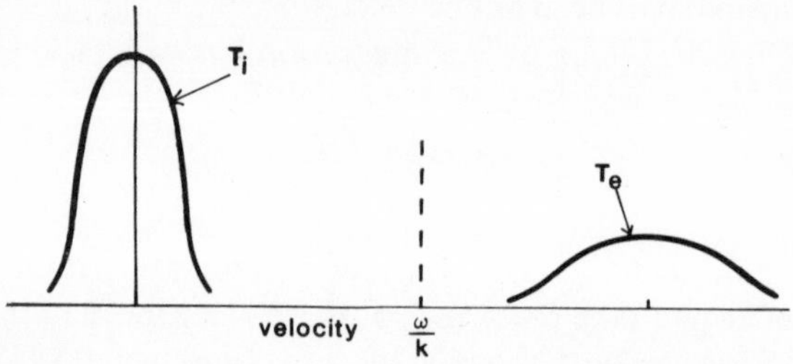

Figure 5.12 Distribution functions for the two-stream instability.

and predicts instability for appropriate values of $\boldsymbol{k}\cdot\boldsymbol{U}$. This instability, the two-stream or Buneman instability, does not rely on resonant wave–particle interaction, but is driven by the entire distribution of streaming electrons. As the electron drift velocity increases there is a transition from the ion sound to this type of instability.

For waves in a magnetic field there are many possible instabilities, so that it is not possible to give any comprehensive account here and we shall restrict our attention to a fairly simple illustrative example. Rather than try to analyse the full electromagnetic dispersion relation, we shall look only at a longitudinal, electrostatic perturbation. Instabilities of this type tend to be more important than electromagnetic instabilities, particularly when the plasma pressure is small compared to the magnetic pressure. The dispersion relation for such waves is obtained from the dielectric tensor as $k_i \varepsilon_{ij} k_j = 0$, that is

$$K(k,\omega) = 1 + \sum_s \frac{q_s^2}{\varepsilon_0 m_s k^2} \sum_{n=-\infty}^{\infty} \int \mathrm{d}^3 v \frac{J_n^2\left(\dfrac{k_\perp v_\perp}{\Omega_s}\right)}{\omega - k_z v_z - n\Omega_s} \times \left[\frac{n\Omega_s}{v_\perp}\frac{\partial f_{0s}}{\partial v_\perp} + k_z \frac{\partial f_{0s}}{\partial v_z}\right] = 0. \tag{5.47}$$

In obtaining instabilities for which $\omega = \omega_r + i\gamma$ with $\gamma \ll \omega_r$ the following procedure is generally adopted. We write

$$K(k,\omega) = K_r(k,\omega) + iK_i(k,\omega) \tag{5.48}$$

where, in most cases, $|K_i| \ll |K_r|$. Then we can obtain a good approximation to ω_r by solving $K_r(k,\omega_r) = 0$, then expand (5.48) to first order to get

$$K(k,\omega_r + i\gamma) \approx K_r(k,\omega_r) + i\gamma \frac{\partial K_r}{\partial \omega_r} + iK_i(k,\omega_r).$$

equating which to zero gives

$$\gamma \approx -\frac{K_i(k,\omega_r)}{\dfrac{\partial K_r}{\partial \omega_r}(k,\omega_r)} \tag{5.49}$$

We shall now use this to look at the Harris instability, which occurs for frequencies near the ion cyclotron harmonics in a plasma where the parallel and perpendicular temperatures are different, that is the distribution functions take the form

$$f_0^s = \frac{n_0 m_s^{3/2}}{(2\pi T_{s\perp}^2 T_{s\parallel})^{3/2}} \exp\left(-\frac{m_s v_\perp^2}{2T_{s\perp}} - \frac{m_s v_\perp^2}{2T_{s\parallel}}\right).$$

The real part of (5.47) is obtained by taking the principal part of the integral and if $k_\perp v_\perp$ is assumed to be small over the important part of the range of integration in the electron term, then the dominant term in the summation is that with $n=0$ and $J_0^2(k_\perp v_\perp/\Omega_s) \approx 1$. Integration by parts, and neglect of the ion term, which is smaller by a factor m_e/m_i, gives us

$$K_r(k, \omega) \approx 1 - \frac{\omega_{pe}^2}{\omega^2}\frac{k_z^2}{k^2},$$

where ω has been assumed much greater than $k_z v_z$, and k_z is non-zero. Thus

$$\omega_r \approx \omega_{pe}\frac{k_z}{k}$$

The imaginary part of (5.47) is a result of the resonant denominator in the integral, and is given by

$$K_i(k, \omega) = -\pi \sum_s \frac{q_s^2}{\varepsilon_0 m_s k^2} \sum_{n=-\infty}^{\infty} \int \mathrm{d}^3 v J_n^2\left(\frac{k_\perp v_\perp}{\Omega_s}\right)$$

$$\times\, \delta(\omega - k_z v_z - n\Omega_s)\left[\frac{n\Omega_s}{v_\perp}\frac{\partial f_{0s}}{\partial v_\perp} + k_z \frac{\partial f_{0s}}{\partial v_z}\right].$$

Substituting the two-temperature distribution in this and using the identity

$$\int_0^\infty x J_n^2(sx)\,\mathrm{e}^{-x^2}\,\mathrm{d}x = \tfrac{1}{2}\mathrm{e}^{-s^2/2} I_n\left(\frac{s^2}{2}\right).$$

we obtain from (5.49)

$$\gamma = \frac{\pi}{2}\omega_r \sum_s \frac{m_s \omega_{ps}^2}{2k^2 T_{s\parallel}} \sum_{n=-\infty}^{\infty} \mathrm{e}^{-k_\perp^2 \rho_s^2/2} I_n\left(\frac{k_\perp^2 \rho_s^2}{2}\right)$$

$$\times \left\{ Z_i'\left(\frac{\omega_r - n\Omega_s}{k_z \alpha_s}\right) - \frac{2n\Omega_s}{k_z \alpha_s}\left(\frac{T_{s\parallel}}{T_{s\perp}}\right) Z_i\left(\frac{\omega_r - n\Omega_s}{k_z \alpha_s^2}\right)\right\} \qquad (5.50)$$

with

$$\alpha_s = \left(\frac{2T_{s\parallel}}{m_s}\right)^{1/2}, \; \rho_s = \frac{1}{\Omega_s}\left(\frac{2T_{s\perp}}{m_s}\right)^{1/2}$$

and Z_i the imaginary part of the plasma dispersion function given by

$$Z_i(z) = \pi^{1/2}\,\mathrm{e}^{-z^2}.$$

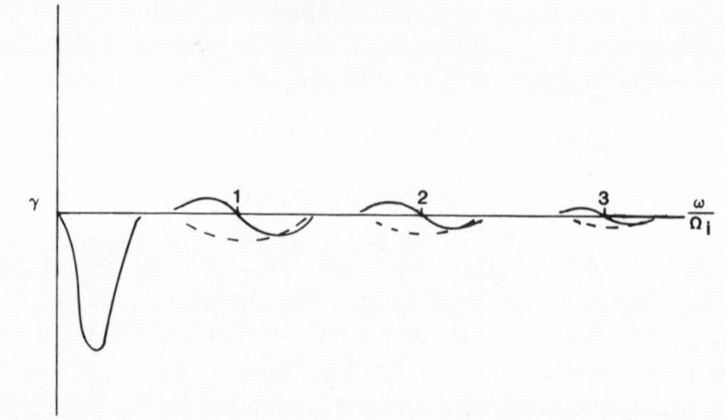

Figure 5.13 Contributions to γ from the terms in (5.48). The negative peak near the origin comes from the electrons, the contributions in full lines near the harmonics from the Z_i' part of the ion term and that in dotted lines from the Z_i part.

For an isotropic distribution it can be verified that the result of (5.50) is negative, corresponding to a damped wave. If, however, we consider the various terms in (5.50) in the general case, we obtain the result represented in Figure 5.13.

If $T_{i\parallel}/T_{i\perp}$ is sufficiently small, the term shown by the dotted lines is small and γ is positive for frequencies just below each of the ion cyclotron harmonics.

The above example represents just one class of distribution function amongst almost endless possibilities for non-Maxwellian distributions which can produce instabilities over a wide range of frequencies. However, it does give some flavour of the sort of analysis used in discussing instabilities in a magnetized plasma. As we saw in Chapter 2, the effect of inhomogeneities in such a plasma is to produce particle drifts and the resulting non-thermal equilibrium distributions produce various drift instabilities. The existence of these means that it is almost impossible to eliminate microinstabilities in any magnetically confined plasma, and one must simply hope that their existence does not affect the confinement too drastically.

5.10 Absolute and convective instabilities

In the foregoing analysis of instabilities we have simply looked at a single Fourier mode of wavenumber k and asked whether the corresponding frequency has an imaginary part corresponding to wave growth. In practice, however, it is important to distinguish two types of unstable behaviour arising from a localized disturbance rather than a sine wave. These two types of behaviour are illustrated in Figure 5.14. In an absolute instability the amplitude grows in time at any fixed point in space, while in a convective

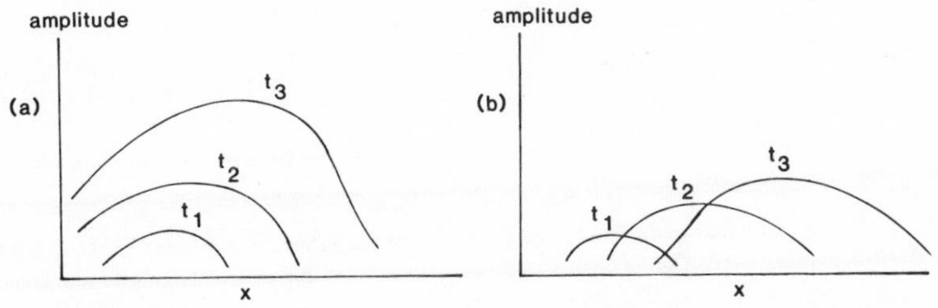

Figure 5.14 Absolute (*a*) and convective (*b*) instabilities ($t_3 > t_2 > t_1$).

instability a disturbance is amplified as it propagates, but at any fixed point it grows and then decays again as the disturbance passes.

This distinction is particularly important in inhomogeneous systems where a wave may only be unstable in some finite region. If the instability is convective then a disturbance only grows as it passes through the unstable region and its growth is limited. To distinguish between these two possibilities it is necessary to look at the initial value problem with a localized initial disturbance and see if its subsequent behaviour is of the type represented by Figure 5.14 (*a*) or (*b*). We shall not pursue the details of such analysis, but simply present the results and justify them as far as possible on a non-rigorous intuitive basis.

For a convective instability we might expect that if we solved the dispersion relation with ω real, then k would be imaginary so as to give the spatial growth of the perturbation. This is indeed the case, but care must be taken to distinguish a true instability from an evanescent wave. For example if we consider the familiar phenomenon of total internal reflection of a wave at an interface, then the solutions in the material of lower dielectric constant are either exponentially decaying or growing away from the interface. One chooses the decaying solution, and there is no question of the growing solution representing an instability. It would occur only if there were a source in the material of lower dielectric constant sending energy towards the interface. In the more complicated plasma medium it may not be so obvious when solutions with a real exponential part do or do not correspond to instability. To decide, it is necessary to consider an initial value problem for a localized perturbation and Laplace transform in time. In inverting the transforms the behaviour depends upon the behaviour of the dispersion relation as a function of the two complex variables k and ω (assuming the direction of $\boldsymbol{k}$ to be fixed).

To decide whether a solution with real ω and imaginary k corresponds to instability the procedure is to take ω first with a large positive imaginary part, corresponding to growth in time of the wave, then let this imaginary part tend to zero. If in the process the imaginary part of k changes sign then the system is convectively unstable. The physical reason can be seen with reference to Figure 5.15. In this diagram (*a*) represents the wave behaviour with an

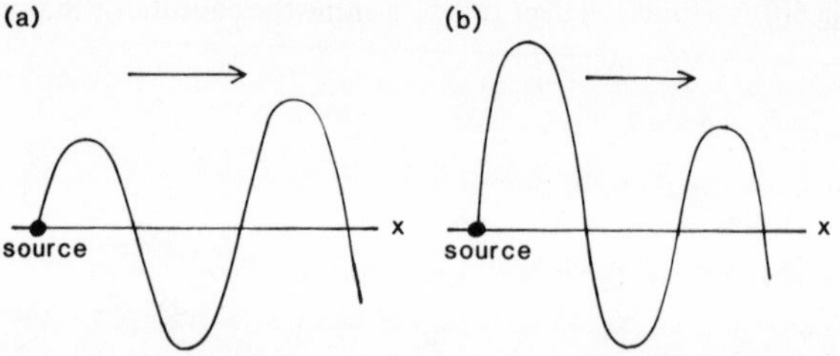

Figure 5.15 Convective instability—(*a*) oscillatory source, (*b*) source with growing amplitude.

oscillatory source showing growth in amplitude as the wave propagates away from the source.

Part (*b*) of the diagram shows what happens if the source has an exponential growth in amplitude. If this is sufficiently rapid the amplitude close to the source can be made larger than that far from the source, the latter having grown because of the convective instability, but from a much lower starting level. In this way the spatial gradient of the amplitude can be reversed, corresponding to the reversal of the imaginary part of k.

An absolute instability occurs if, as we go through the above process, two roots k_1 and k_2 approach each other and coalesce before the imaginary part of ω goes to zero. In this case the imaginary part of ω cannot be taken smoothly to zero and there is an exponential growth in time.

Problems

5.1 By means of successive integration by parts of the integral in (5.30) obtain the expansions of $Z(\zeta)$ for small and large real ζ.

5.2 Show that the damping rate of ion sound waves is approximately

$$\gamma = \frac{\pi\omega_r}{2}\sum_s \frac{\omega_{ps}^2}{(k^2 + 1/\lambda_D^2)} F'_{0s}\left(\frac{\omega_r}{k}\right),$$

with ω_r given by (5.32).

5.3 Suppose that one or more constants of motion can be found for charged particles in steady electric and magnetic fields. (The total energy will be one such). Show that a time-independent solution of the Vlasov equation for each species is given by an arbitrary function of these constants of motion.

5.4 Obtain the dispersion relation (5.46) and show that it predicts instability if

$$(\boldsymbol{k}\cdot\boldsymbol{U})^2 < \left(1 + \left(\frac{m_e}{m_i}\right)^{1/3}\right)^3 \omega_{pe}^2.$$

Hint: Sketch (5.46) as a function of ω and determine the conditions that it have only two real roots.

5.5 Obtain the result of question 5.4 using the Nyquist diagram technique. In Figure 5.10 (*a*) take the straight part of the contour to be just above the real axis.

5.6 Show that as $k \to 0$ the frequency of a Bernstein mode tends to $n\Omega$ from below if this frequency is less than the upper hybrid frequency, and from above if it is greater than the upper hybrid frequency.

6 Nonlinear effects

6.1 Introduction

In Chapters 3 and 5 we have considered a number of instabilities, characterized by a solution in which small perturbations to an equilibrium are predicted to grow exponentially in time. Such growing perturbations will, of course, not remain small and eventually the nonlinear terms which are neglected in our previous analysis must become important. The object of this chapter is to examine various theories dealing with the nonlinear behaviour of the plasma.

We look first at quasilinear theory, one of the oldest and most widely used nonlinear methods, which deals with the way in which waves may react back on the equilibrium distribution function and change it in such a way as to stabilize the equilibrium. Then we turn to wave–wave interactions. Waves which propagate independently, obeying a simple principle of superposition, in the linear approximation can interact through the nonlinear terms, and if appropriate frequency and wavelength matching conditions are satisfied give rise to a rich variety of effects. One such effect of particular importance is parametric instability, when a large-amplitude incident wave excites other plasma modes of different frequencies.

Finally we shall say something about computational methods, which are of great importance in the theoretical analysis of nonlinear behaviour since nonlinear analytical methods are limited in scope and difficult to apply to realistic problems. Two main methods will be discussed, fluid and particle codes. The first aims to implement numerical schemes for the solution of fluid equations and involves finite differencing of these partial differential equations, or possibly other numerical methods like the finite element technique. Particle codes are aimed at examining the microscopic behaviour of the plasma which, analytically, would be looked at using the Vlasov equation. The general idea is to simulate the plasma behaviour by following the orbits of a large number of charged particles moving in self-consistent electric and magnetic fields.

6.2 Quasilinear theory

In this theory the particle distribution function is split into a slowly varying average part plus a rapidly fluctuating part produced by the wave. We shall consider its simplest application, to electrostatic waves in an unmagnetized plasma at high frequencies where only the electron motion is important. Writing the distribution function for electrons as

$$f(\boldsymbol{r}, \boldsymbol{v}, t) = f_0(\boldsymbol{v}, t) + f_1(\boldsymbol{r}, \boldsymbol{v}, t),$$

where f_0 is an averaged distribution function varying over a much longer time scale than the wave period on which f_1 varies, and substituting this in the Vlasov equation gives

$$\frac{\partial f_0}{\partial t} + \frac{\partial f_1}{\partial t} + \boldsymbol{v}\cdot\frac{\partial f_1}{\partial \boldsymbol{r}} - \frac{e}{m}\boldsymbol{E}\cdot\frac{\partial f_0}{\partial \boldsymbol{v}} - \frac{e}{m}\boldsymbol{E}\cdot\frac{\partial f_1}{\partial \boldsymbol{v}} = 0. \tag{6.1}$$

The electric field, the average of which is taken to be zero, is found from

$$\boldsymbol{\nabla}\cdot\boldsymbol{E} = -\frac{e}{\varepsilon_0}\int f_1\,\mathrm{d}^3 v. \tag{6.2}$$

If we take the average of (6.1) over the fast time scale we get

$$\frac{\partial f_0}{\partial t} - \frac{e}{m}\left\langle \boldsymbol{E}\cdot\frac{\partial f_1}{\partial \boldsymbol{v}}\right\rangle = 0 \tag{6.3}$$

the angular brackets denoting the average. Terms linear in the rapidly fluctuating quantities disappear when the average is taken. Subtracting (6.1) from (6.3) gives

$$\frac{\partial f_1}{\partial t} + \boldsymbol{v}\cdot\frac{\partial f_1}{\partial \boldsymbol{r}} - \frac{e}{m}\boldsymbol{E}\cdot\frac{\partial f_0}{\partial \boldsymbol{r}} = \frac{e}{m}\left(\boldsymbol{E}\cdot\frac{\partial f_1}{\partial \boldsymbol{v}} - \left\langle \boldsymbol{E}\cdot\frac{\partial f_1}{\partial \boldsymbol{v}}\right\rangle\right). \tag{6.4}$$

The essential approximation of quasilinear theory is the neglect of the right-hand side of (6.4), so that f_1 is just described by the usual linearized Vlasov equation, the only difference being that f_0 is no longer constant, but varies on the slow time scale according to (6.3). In this approximation interactions between wave modes are being neglected and all that is being considered is the reaction of the waves back on to the averaged distribution function.

To solve (6.4) we make the further assumption that on the short time scale the time dependence of f_0 can be neglected, so that we just solve (6.4), together with (6.2) as in the last chapter. Transient effects arising from their solution as an initial value problem will be neglected, so that we have perturbations going as $\exp(i\boldsymbol{k}\cdot\boldsymbol{r} - i\omega t)$, with ω and $\boldsymbol{k}$ connected by the usual dispersion relation. Over the long time scale, as f_0 changes, ω will change for a given k. Thus when we look at the oscillations of the plasma on a short time scale we simply see waves obeying the usual dispersion relation. Over the longer time scale there is a gradual change in the dispersion properties and hence of the wave frequency and, more importantly, growth rate.

If we suppose that the electric field consists of a superposition of Fourier components $\boldsymbol{E}_k$ corresponding to the wavenumber $\boldsymbol{k}$, then from (6.4)

$$f_1 = i\frac{e}{m}\sum_k \boldsymbol{E}_k \cdot \frac{\partial f_0}{\partial \boldsymbol{v}} \Bigg/ (\omega - \boldsymbol{k}\cdot\boldsymbol{v})$$

substituting which in (6.3) gives, in tensor notation,

$$\frac{\partial f_0}{\partial t} = \frac{\partial}{\partial v_i}\mathscr{D}_{ij}\frac{\partial f_0}{\partial v_j} \tag{6.5}$$

with the coefficient $\mathscr{D}_{ij}$ given by

$$\mathscr{D}_{ij} = \frac{1}{2}\frac{e^2}{m^2} Im \sum_k \frac{E_{ki}E^*_{kj}}{\omega - \boldsymbol{k}\cdot\boldsymbol{v}} = \frac{1}{2}\frac{e^2}{m^2} Im \sum_k \frac{|\boldsymbol{E}_k|^2 \hat{k}_i \hat{k}_j}{\omega - \boldsymbol{k}\cdot\boldsymbol{v}}. \tag{6.6}$$

The last equality in (6.6) depends on the fact that the waves are longitudinal. Products of different Fourier components are rapidly oscillating and average to zero.

To complete the description we must consider the time-dependence of the electric field components. If γ is the growth rate of the wave with wavenumber $\boldsymbol{k}$ then the short time scale solution has $|\boldsymbol{E}_k|^2$ proportional to $e^{2\gamma t}$. Over the long time scale, on which γ has to be taken to be time-dependent, this becomes $\exp(2\int\gamma(t)\,dt)$. Usually this behaviour of $|\boldsymbol{E}_k|^2$ is expressed by noting that it is the solution of the differential equation

$$\frac{\partial}{\partial t}|\boldsymbol{E}_k|^2 = 2\gamma(\boldsymbol{k}, t)|\boldsymbol{E}_k|^2. \tag{6.7}$$

The end result of the theory is therefore the set of coupled equations (6.5) and (6.7) which determine the evolution of the average distribution function and the electric field spectrum on the long time scale.

Before examining the behaviour of these equations let us first show that they conserve energy, always a desirable property of any physical theory. The rate of change of the energy associated with the average particle distribution is

$$\int \tfrac{1}{2}mv^2 \frac{\partial f_0}{\partial t}\,d^3v,$$

which becomes, after using (6.5) and carrying out an integration by parts,

$$\sum_k \frac{1}{2}\frac{e^2}{m^2} Im \int \frac{(\hat{\boldsymbol{k}}\cdot\boldsymbol{v})\left(\hat{\boldsymbol{k}}\cdot\dfrac{\partial f_0}{\partial \boldsymbol{v}}\right)}{\omega - \boldsymbol{k}\cdot\boldsymbol{v}}\,d^3v\,|\boldsymbol{E}_k|^2. \tag{6.8}$$

The current associated with the wave mode of wavenumber k is

$$\boldsymbol{J}_k = e\int f_{1k}\,d^3v$$

$$= i\frac{e^2}{m}\int \frac{\boldsymbol{E}_k\cdot\dfrac{\partial f_0}{\partial \boldsymbol{v}}}{\omega - \boldsymbol{k}\cdot\boldsymbol{v}}\,\mathrm{d}^3v, \tag{6.9}$$

and the rate of change of the average energy associated with this mode is (cf. section 4.4)

$$\frac{\mathrm{d}W_k}{\mathrm{d}t} = \tfrac{1}{2}Re(\boldsymbol{E}_{\boldsymbol{k}}^*\cdot\boldsymbol{J}_k).$$

which, using (6.9) and the longitudinal nature of the electric fields, can be shown to be

$$-\frac{1}{2}\frac{e^2}{m^2}Im\int \frac{(\hat{\boldsymbol{k}}\cdot\boldsymbol{v})\left(\hat{\boldsymbol{k}}\cdot\dfrac{\partial f_0}{\partial \boldsymbol{v}}\right)}{\omega - \boldsymbol{k}\cdot\boldsymbol{v}}\,\mathrm{d}^3v\,|\boldsymbol{E}_k|^2.$$

Comparing this with (6.8) shows that the change in wave energy is exactly balanced by the change in the average particle kinetic energy associated with the average distribution function.

Writing $\omega = \omega_r + i\gamma$ gives (6.6) in the form

$$\mathscr{D}_{ij} = \frac{1}{2}\frac{e^2}{m^2}\sum_{\boldsymbol{k}}\frac{\gamma|\boldsymbol{E}_k|^2\hat{k}_i\hat{k}_j}{(\omega_r - \boldsymbol{k}\cdot\boldsymbol{v})^2 + \gamma^2}. \tag{6.10}$$

This holds for $\gamma > 0$, that is growing waves, but for $\gamma < 0$ there is the usual problem that the solutions we have used hold only in the upper half plane. In integrals like (6.8) and (6.9) the usual method of analytic continuation, discussed in detail in Chapter 5, can be used, but it is not obvious how to treat (6.10). For small γ the relation

$$Im\left(\frac{1}{\omega - \boldsymbol{k}\cdot\boldsymbol{v}}\right) \rightarrow \pi\delta(\omega - \boldsymbol{k}\cdot\boldsymbol{v}) \quad \text{as} \quad \gamma \rightarrow 0 \tag{6.11}$$

is often used, giving a meaningful expression for $\mathscr{D}_{ij}$ if the sum over a finite wavenumber spectrum is replaced by an integral over a continuous spectrum. The use of quasilinear theory is best restricted to these regions of the wave-number spectrum which are unstable or whose damping rate is so small that this last approximation is reasonable. These are usually the regions of most interest, so this is not an unduly severe restriction.

Equation 6.5 takes the form of a diffusion equation in velocity space and as with all diffusion equations its solution is such as to smooth out gradients as time goes on. From the form (6.10) of the diffusion coefficient it can be seen that the contribution from the mode $\boldsymbol{k}$ is greatest for those particles for which $\omega - \boldsymbol{k}\cdot\boldsymbol{v} \approx 0$, that is the particles in resonance with the wave as discussed in Chapter 5, a result which is not particularly surprising. If the limiting form for small growth rate (6.11) is used the wave affects only particles for which

$\omega - \boldsymbol{k} \cdot \boldsymbol{v}$ is exactly zero. The qualitative behaviour of the solution of (6.5) is therefore expected to be a flattening of f_0 in that region of velocity space where particles are in resonance with unstable waves. Since the instability is generally driven by a gradient in velocity space the effect is to reduce the growth rate and eventually, perhaps, stabilize the system.

A simple example is the 'bump on tail' instability discussed in section 5.9, where it was shown that a one-dimensional distribution function like that shown in Figure 6.1(*a*) is unstable to waves with phase velocities in the range where $F'_0(u)$ is positive. The final state according to quasilinear theory is shown in Figure 6.1(*b*) where a spectrum of unstable waves has grown in the appropriate region, flattening the distribution function there and eventually leading to a steady state where the wave growth rate goes to zero and at the same time the right-hand side of (6.5) vanishes because the gradient of the distribution function is zero over the important part of its range.

Quasilinear theory can be generalized to electromagnetic waves and to magnetized plasmas. In the latter case the diffusion coefficient is greatest for particles with $\omega - k_z v_z - n\Omega \approx 0$, for the reasons discussed in Chapter 5 where we dealt with cyclotron resonances. Always, however, the qualitative effect is of a flattening of the distribution function in the resonant regions of velocity space, and a consequent stabilization of instabilities associated with velocity space gradients. The diffusion behaviour of the velocity distribution function gives some clue as to the circumstances under which the theory would be expected to be a good approximation. Essentially there should be a broad spectrum of waves so that a particle undergoes a random scattering process under the influence of the waves, rather than the organized motion which might be expected in a regular large-amplitude sinusoidal wave.

In waves of sufficiently large amplitude a significant number of particles may be trapped in the potential troughs of the waves, as described in section 5.3. This means that particles in some range of velocities around those for which $\omega = \boldsymbol{k} \cdot \boldsymbol{v}$ interact strongly with the waves, the range increasing with wave amplitude. Modifications of quasilinear theory have been developed to take this into account, through an amplitude-dependent broadening of the resonance function in 6.10.

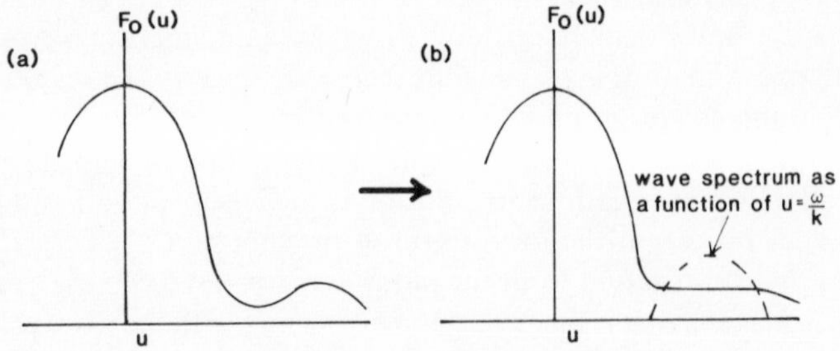

Figure 6.1 Evolution of one-dimensional 'bump on tail' instability according to quasilinear theory.

6.3 Wave–wave interactions

Now, instead of looking at the effect of waves on the average behaviour of the plasma, we turn to the question of how different wave modes, which propagate independently in linear theory, interact through nonlinear terms. Some quite simple considerations will tell us under what conditions we may expect an important effect. First, let us recall that a wave which in linear theory is taken to behave like $a_0 \mathrm{e}^{i\boldsymbol{k}\cdot\boldsymbol{r}-i\omega t}$ is actually described by a physical amplitude which is the real part of this, that is

$$\tfrac{1}{2}(a_0 \mathrm{e}^{i\boldsymbol{k}\cdot\boldsymbol{r}-i\omega t} + a_0^* \mathrm{e}^{-i\boldsymbol{k}\cdot\boldsymbol{r}+i\omega t}).$$

This is important where products of amplitudes are concerned.

Now suppose that there are two waves which go as $\exp(i\boldsymbol{k}_1\cdot\boldsymbol{r} - i\omega_1 t)$ and $\exp(i\boldsymbol{k}_2\cdot\boldsymbol{r} - i\omega_2 t)$ in the linear approximation. Then, taking the above remarks into account, it can be seen that a nonlinear term containing the product of the two wave amplitudes contains terms going as $\exp\{i(\boldsymbol{k}_1 + \boldsymbol{k}_2)\cdot\boldsymbol{r} - i(\omega_1 + \omega_2)t\}$, $\exp\{i(\boldsymbol{k}_1 - \boldsymbol{k}_2)\cdot\boldsymbol{r} - i(\omega_1 - \omega_2)t\}$ and their complex conjugates. These terms act as driving terms at the sum and difference frequencies and wavenumbers. Generally, for wave amplitudes which are not very large, such driving terms are quite small and would only be expected to drive up small second-order perturbations at the appropriate frequencies. The exception occurs when the sum or difference frequencies and wavenumbers themselves belong to one of the linear modes of the plasma. Thus we are led to the idea of a resonant triad, a group of three waves whose frequencies and wavenumbers satisfy

$$\omega_1 = \omega_2 + \omega_3 \tag{6.12}$$

and

$$\boldsymbol{k}_1 = \boldsymbol{k}_2 + \boldsymbol{k}_3. \tag{6.13}$$

(The relations can always be written like this, with a plus sign, by relabelling the modes if necessary). In such a group the nonlinear terms involving any two waves contain terms which are in resonance with the third wave, and so there is a strong interaction amongst the waves. If terms of higher order in the wave amplitude are considered then four and more wave interactions are possible, requiring similar matching conditions. However, if resonant triads exist they will give the most important effect at reasonably small wave amplitudes.

In a plasma there is such a variety of wave modes that there is no difficulty in finding examples of resonant triads. In order to illustrate the general principles involved in investigating their behaviour we shall consider, instead of equations describing plasma waves, a hypothetical system in which a wave amplitude ϕ obeys the equation

$$\frac{\partial^2 \phi}{\partial t^2} = \nabla^2\phi - \nabla^2(\nabla^2\phi) + \phi^2. \tag{6.14}$$

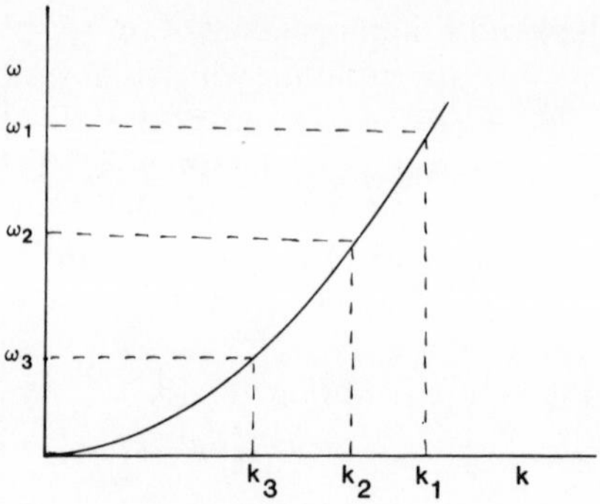

Figure 6.2 Linear dispersion curve corresponding to equation 6.14.

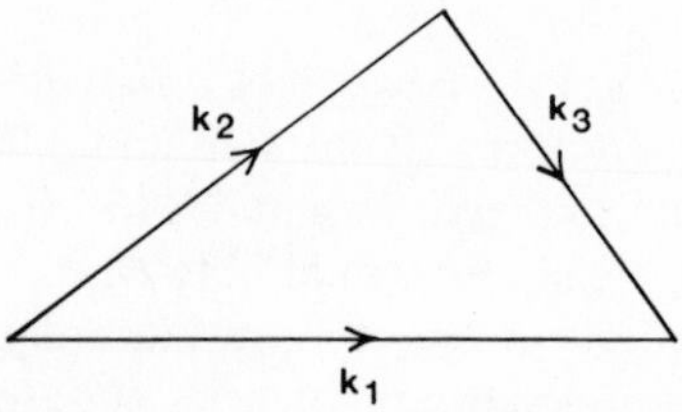

Figure 6.3 Wavenumber matching condition.

If this is linearized we obtain the dispersion relation

$$\omega^2 = k^2 + k^4$$

which is shown in Figure 6.2.

It is clear that

$$k_1 < k_2 + k_3$$

so that it is possible to choose vectors $\boldsymbol{k}_1, \boldsymbol{k}_2$ and $\boldsymbol{k}_3$ such that

$$\boldsymbol{k}_1 = \boldsymbol{k}_2 + \boldsymbol{k}_3$$

as illustrated in Figure 6.3.

Let us suppose then that we have three waves, which, when the nonlinear term in (6.14) is neglected, correspond to solutions $\phi_i e^{i\boldsymbol{k}_i \cdot \boldsymbol{r} - i\omega_i t} (i = 1, 2, 3)$. In this approximation ϕ_i is constant.

For amplitudes which are not too large we can consider a weakly nonlinear theory, in which the wave modes are assumed to retain their basic identity, but are no longer completely independent. However, the time scale over which their amplitudes change owing to the interchange of energy amongst them is taken to be long compared to the wave periods. Assuming the wave amplitudes to be spatially homogeneous, we represent them as $\phi_i(t)$ $e^{ik_i\cdot r - i\omega_i t}(i = 1, 2, 3)$, with $\phi_i(t)$ an amplitude which is now slowly varying instead of constant.

With the total ϕ a superposition of these waves, the term in ϕ^2 which drives ϕ_1 is

$$\tfrac{1}{4}\phi_2\phi_3 e^{i(k_2+k_3)\cdot r - i(\omega_2+\omega_3)t} + \text{complex conjugate} = Re(\tfrac{1}{2}\phi_2\phi_3 e^{ik_1\cdot r - i\omega_1 t}).$$

Now

$$\frac{\partial^2}{\partial t^2}(\phi_1(t)e^{ik_1\cdot r - i\omega_1 t}) = \left(-\omega^2\phi_1 - 2i\omega_1\frac{d\phi_1}{dt} + \frac{d^2\phi_1}{dt^2}\right)e^{ik_1\cdot r - i\omega_1 t}$$

so, if we substitute this in (6.14) and neglect $d^2\phi_1/dt^2$ on the basis that ϕ_1 is slowly varying, we obtain

$$-2i\omega_1\frac{d\phi_1}{dt} = \tfrac{1}{2}\phi_2\phi_3$$

or

$$\frac{d\phi_1}{dt} = \frac{i}{4\omega_1}\phi_2\phi_3. \tag{6.15}$$

In a similar way we obtain

$$\frac{d\phi_2}{dt} = \frac{i}{4\omega_2}\phi_1\phi_3^*$$

$$\frac{d\phi_3}{dt} = \frac{i}{4\omega_3}\phi_1\phi_2^*.$$

A more symmetric form of (6.15) and its companions can be obtained if we define $C_i = \omega_i^{1/2}\phi_i$ so that we have

$$\begin{aligned}\frac{dC_1}{dt} &= i\lambda C_2C_3\\ \frac{dC_2}{dt} &= i\lambda C_1C_3^*\\ \frac{dC_3}{dt} &= i\lambda C_1C_2^*,\end{aligned} \tag{6.16}$$

with

$$\lambda = \tfrac{1}{4}(\omega_1\omega_2\omega_3)^{1/2}.$$

The calculation for plasma waves proceeds in the same way, being straightforward in principle but rather tedious in detail since we start with a system of equations rather than a single rather simple equation. A simple rescaling of the wave amplitudes will always reduce the equations to a symmetric form like (6.16) where all the coupling coefficients are equal in magnitude.

From (6.16) and (6.12) it follows that

$$\frac{\mathrm{d}}{\mathrm{d}t}\{\omega_1|C_1|^2+\omega_2|C_2|^2+\omega_3|C_3|^2\}=0 \tag{6.17}$$

If the amplitudes C_1, C_2, C_3 which reduce the equations to the form (6.16) are calculated for any set of plasma waves (or waves in any other system for that matter) it will be found that equation (6.17) can be interpreted as an equation for conservation of energy, with $\omega_i|C_i|^2$ being, to within a constant multiple, the energy density of the wave (as discussed in section 4.4). We assume here positive energy waves, there being an appropriate change of sign if one of the modes is a negative energy wave.

It is also a straightforward consequence of (6.16) that

$$\frac{\mathrm{d}}{\mathrm{d}t}(|C_1|^2+|C_2|^2)=\frac{\mathrm{d}}{\mathrm{d}t}(|C_1|^2+|C_3|^2)=\frac{\mathrm{d}}{\mathrm{d}t}(|C_2|^2-|C_3|^2) \tag{6.18}$$

These are known as the Manley–Rowe relations. A considerable effort has been expended over the behaviour of (6.16) and its generalizations, some of which will be mentioned below. It is possible to find an exact solution of (6.16), in terms of elliptic functions, but we shall simply give a brief account of the qualitative features of the solution. Let us suppose that C_1 begins at a much larger amplitude than C_2 or C_3. Then initially C_2 and C_3 grow at the expense of C_1. It is, however, clear from (6.17) and (6.18) that the solutions are bounded and eventually this growth ceases and energy flows from C_2 and C_3 back to C_1. In general solutions are periodic in nature.

Various generalizations to (6.16) are possible. For instance weak wave damping may be included by having $\mathrm{d}C_i/\mathrm{d}t - \nu_i C_i$ on the left-hand sides, with ν_i the damping rate of the ith wave in the absence of nonlinear effects. An important generalization is the interaction of wave packets in which the amplitudes vary in space, in which case (6.16) becomes

$$\frac{\partial C_1}{\partial t}+(\boldsymbol{v}_1\cdot\nabla)C_1=i\lambda C_2C_3,$$

and so on, $\boldsymbol{v}_i$ being the group velocity of the ith wave. Some analytic progress is sometimes possible with these more general equations, but often the problem is best tackled by numerical means.

6.4 Parametric instabilities

In general terms a parametric instability is one in which some mode of a system is driven unstable by an externally imposed oscillation at a different frequency, and investigation of such instabilities has a history stretching back for more than a century. In plasma physics the externally imposed oscillation is generally a large-amplitude wave excited by an external source which, by means of a coupling process like that discussed in section 6.2, excites other waves in the plasma. Parametric instabilities of this sort have been investigated intensively, to a large extent because of their interest in the subject of laser–plasma interactions, where they provide a means by which energy in the incident electromagnetic wave can be fed into other types of wave.

If we suppose that there are two waves, which together with the incident, or 'pump', wave form a resonant triad with the pump wave having the highest frequency, then from (6.16) their amplitudes are described by

$$\frac{\mathrm{d}C_2}{\mathrm{d}t} - \nu_2 C_2 = i\lambda C_1 C_3^*$$

$$\frac{\mathrm{d}C_3}{\mathrm{d}t} - \nu_3 C_3 = i\lambda C_1 C_2^*, \tag{6.19}$$

where ν_2 and ν_3 are damping coefficients which we now wish to take into account. The amplitude of the pump wave C_1 may be assumed constant while the other waves are of small amplitude, and with this approximation the pair of equations (6.19) has a solution going as e^{pt} where

$$(p - \nu_2)(p - \nu_3) = \lambda^2 |C_1|^2.$$

This predicts positive p, giving growth of the two decay waves, if

$$\lambda^2 |C_1|^2 > \nu_2 \nu_3. \tag{6.20}$$

Equation 6.20 is an important result showing one of the characteristic features of a parametric instability, namely that the instability occurs only if the intensity of the pump wave is above a certain threshold value which is proportional to the product of the damping coefficients of the two decay waves. This threshold behaviour is one of the ways in which effects resulting from parametric instabilities may be identified in experiments.

Equations 6.19 assume perfect wavenumber and frequency matching. If we suppose that there is a frequency mismatch, so that

$$\omega_1 - \omega_2 - \omega_3 = \Delta\omega,$$

then (6.19) becomes

$$\frac{\mathrm{d}C_2}{\mathrm{d}t} - \nu_2 C_2 = i\lambda C_1 C_3^* \,\mathrm{e}^{i\Delta\omega t}$$

$$\frac{\mathrm{d}C_3}{\mathrm{d}t} - \nu_3 C_3 = i\lambda C_1 C_2^* \,\mathrm{e}^{i\Delta\omega t}.$$

If we let

$$\tilde{C}_2 = C_2 \,\mathrm{e}^{-i\Delta\omega t/2}, \quad \tilde{C}_3 = C_3 \,\mathrm{e}^{-i\Delta\omega t/2},$$

we get

$$\frac{\mathrm{d}\tilde{C}_2}{\mathrm{d}t} - \left(i\frac{\Delta\omega}{2} + \nu_2 \right) \tilde{C}_2 = i\lambda C_1 \tilde{C}_3^*$$

$$\frac{d\tilde{C}_3^*}{dt} - \left(-i\frac{\Delta\omega}{2} + \nu_3\right)\tilde{C}_3^* = -i\lambda C_1^* C_2,$$

so that if we once more look for a solution going as e^{pt} we obtain

$$\left[p - \left(i\frac{\Delta\omega}{2} + \nu_2\right)\right]\left[p - \left(-\frac{i\Delta\omega}{2} + \nu_3\right)\right] = \lambda^2 |C_1|^2. \tag{6.21}$$

If we let $p = p_1 + ip_2$ and note that p_1 must go through zero at the threshold value of $|C_1|^2$, we can eliminate p_2 from the real and imaginary parts of (6.21) and obtain the threshold condition

$$\lambda^2 |C_1|^2 = \nu_2 \nu_3 + \Delta\omega^2 \frac{\nu_2 \nu_3}{(\nu_2 + \nu_3)^2}. \tag{6.22}$$

The minimum threshold occurs when there is perfect frequency matching, as might be expected, and the further above this minimum threshold we take the pump intensity, the greater the frequency mismatch which may be tolerated before the instability is quenched.

All of the above has assumed a homogeneous plasma, but in practice, of course, this condition is rarely satisfied, even approximately. As we discussed in Chapter 4, as a wave propagates in an inhomogeneous plasma its wavenumber may be supposed to change, being given by the local dispersion relation. This means that if a pump and two decay waves satisfy the wavenumber matching condition at some point then they no longer do so at other adjacent points, and a wavenumber mismatch appears which tends to reduce the growth of the instability in a similar fashion to a frequency mismatch. In an inhomogeneous system an instability involving a given wave triad may occur only in a restricted region, and the result is to reduce the growth rate and increase the threshold as compared with the homogeneous plasma results. A great deal of work has been done on parametric instabilities in inhomogeneous systems motivated by the relevance of such studies to the theory of laser–plasma interactions. However, this is a complicated subject which it would not be appropriate to pursue in detail here.

Let us turn instead to look at some particular parametric instabilities which may be important in laser–plasma interactions. Any magnetic fields present in laser-irradiated targets are not strong enough to affect greatly the wave propagation and so we shall consider only an unmagnetized plasma. The pump wave is an electromagnetic mode and possible decay modes are, in addition to other electromagnetic waves, the plasma and ion sound waves. Examination of the possible decays shows that the following processes are possible.

e.m. $\rightarrow$ plasma + i.s. (parametric decay instability)

e.m. $\rightarrow$ e.m. + i.s (stimulated Brillouin scattering) e.m. $\rightarrow$ e.m. + plasma (stimulated Raman scattering)

e.m. $\rightarrow$ plasma + plasma (two-plasmon decay)

The first of these was the first such instability to be investigated in detail and came to be known as simply the 'parametric decay instability', though this term may also be applied to the process of parametric instability more generally. The next two instabilities are named from their similarity to processes taking place in the interaction of light with solids or liquids.

Since the plasma wave can only propagate without being very strongly damped at frequencies a little above the plasma frequency, and the ion sound wave is at a much lower frequency, the decay instability occurs only where the plasma frequency is slightly less than the frequency of the incident wave. It is confined to a region just outside the critical surface where these two frequencies become equal. For similar reasons the two-plasmon decay occurs only near the surface at which the plasma frequency is one-half of the incident wave frequency. Since the plasma frequency is proportional to the square root of the density this is the quarter-critical surface. The electromagnetic wave, on the other hand, can propagate at any frequency above the plasma frequency, and so stimulated Brillouin scattering is possible throughout the region where the density is below critical, while stimulated Raman scattering can take place on the low-density side of the quarter-critical surface. These last two instabilities lead to a transfer of part of the energy of the incoming mode into another electromagnetic wave which may be backwards or sideways propagating and so produce a scattering of light from the target instead of absorption. If there is to be effective absorption of the incident laser light it is important that neither of them should grow to a high level.

In this connection we should point out an important consequence of the last of the Manley–Rowe relations (6.18), which is that the energy in the two decay waves is proportional to their frequencies. Thus, in stimulated Brillouin scattering most of the energy goes into the electromagnetic wave, which is at a much higher frequency than the ion sound wave. In principle this instability could lead to scattering of almost all the incident radiation, though in practice the situation does not seem to be so serious.

We have considered only the initial stage of the instability in which the decay waves are small and do not have an appreciable effect on the pump wave. As in the case of any instability the initial exponential growth cannot continue indefinitely and to predict the final state of the plasma requires a much more complicated nonlinear analysis. Various effects may limit the growth of a parametric instability, the most obvious being depletion of the pump as the decay waves begin to extract appreciable amounts of energy from it. Various other effects have been suggested as limiting the amplitude. For instance, the decay waves may themselves become unstable to further decays, so energy is fed into a cascade of modes. Or, if an electrostatic decay wave is driven to large amplitude it may trap and accelerate particles in its potential troughs and so be damped at a rate larger than the linear damping rate. If the effective damping becomes large enough the pump intensity may be below threshold and the instability may be switched off. Particle trapping and acceleration appears to be an important process in many laser–plasma interaction experiments, leading to the production of a tail of high-energy particles in the electron distribution function. If the laser light is being used to compress and heat a target these fast particles are undesirable, since they run ahead and heat the core before the main compression wave has arrived. This makes the compression less effective. It is best to choose parameters where absorption is mainly by electron–ion collisions, rather than through parametric instabilities or resonant absorption (which is discussed in Chapter 4). In experiments on laser–plasma interactions all the possible parametric instabilities and many other effects occur simultaneously and to disentangle the various phenomena in the experimental data is a very difficult task.

6.5 The ponderomotive force

If a charged particle is placed in an electromagnetic wave field of uniform amplitude then it simply oscillates, but if the amplitude is non-uniform the force on it may contain a steady component, which is like a pressure due to the radiation. This force is known as the ponderomotive force and acts to push particles into regions in which the wave field is less intense. It plays an important role in laser–plasma interactions where the very intense wave fields can exert forces strong enough to have a major role in the particle dynamics.

To calculate this force we consider the response of a charged particle in an electromagnetic field in which

$$\boldsymbol{E} = \boldsymbol{E}_0(\boldsymbol{r})\cos(\boldsymbol{k}\cdot\boldsymbol{r} - \omega t) \tag{6.22}$$

so that, from $\nabla \times \boldsymbol{E} = -\partial \boldsymbol{B}/\partial t$,

$$\boldsymbol{B} = \boldsymbol{B}_0(\boldsymbol{r})\sin(\boldsymbol{k}\cdot\boldsymbol{r} - \omega t) + \frac{1}{\omega}\boldsymbol{k} \times \boldsymbol{E}_0 \cos(\boldsymbol{k}\cdot\boldsymbol{r} - \omega t), \tag{6.23}$$

with

$$\boldsymbol{B}_0 = -\frac{1}{\omega}\nabla \times \boldsymbol{E}_0.$$

If the spatial dependence of $\boldsymbol{E}_0$ is neglected, a first approximation to the particle motion is obtained, namely,

$$\boldsymbol{v} = -\frac{q}{m\omega}\boldsymbol{E}_0 \sin(\boldsymbol{k}\cdot\boldsymbol{r} - \omega t)$$

$$\boldsymbol{r} = \boldsymbol{r}_0 - \frac{q}{m\omega^2}\boldsymbol{E}_0 \cos(\boldsymbol{k}\cdot\boldsymbol{r} - \omega t).$$

To obtain a better approximation we substitute this back into the right-hand side of the equation of motion,

$$\frac{\mathrm{d}\boldsymbol{v}}{\mathrm{d}t} = \frac{q}{m}(\boldsymbol{E} + \boldsymbol{v} \times \boldsymbol{B}), \tag{6.24}$$

and use the approximation

$$\boldsymbol{E}_0(\boldsymbol{r}) \approx \boldsymbol{E}_0(\boldsymbol{r}_0) - \frac{q}{m\omega^2}\cos(\boldsymbol{k}\cdot\boldsymbol{r} - \omega t)(\boldsymbol{E}_0\cdot\boldsymbol{\nabla})\boldsymbol{E}_0$$

in the electric field term. In the magnetic force term take $\boldsymbol{E}_0(\boldsymbol{r}) = \boldsymbol{E}_0(\boldsymbol{r}_0)$. If, with these approximations, (6.24) is averaged over the oscillations represented by the sine and cosine terms the result is

$$\left\langle \frac{\mathrm{d}v}{\mathrm{d}t} \right\rangle = \frac{1}{2}\frac{q}{m}\left\{ -\frac{q}{m\omega^2}(\boldsymbol{E}_0\cdot\boldsymbol{\nabla})\boldsymbol{E}_0 - \frac{q}{m\omega^2}\boldsymbol{E}_0 \times (\boldsymbol{\nabla} \times \boldsymbol{E}_0)\right\}$$

$$= -\frac{1}{4}\frac{q^2}{m^2\omega^2}\boldsymbol{\nabla}(E_0^2). \tag{6.25}$$

Thus, on averaging over the rapid oscillations of the field there remains a steady component of force

$$\boldsymbol{F} = -\frac{1}{4}\frac{q^2}{m\omega^2}\boldsymbol{\nabla}E_0^2 \tag{6.26}$$

in the presence of a gradient in the field amplitude. This is the ponderomotive force, the effect of which is to push particles towards regions of lower wave intensity. The physical origin of this force depends on whether the gradient of the field amplitude is parallel or perpendicular to the field itself. In the former case the force arises from the first term on the right-hand side of (6.25) and its physical origin may be explained with reference to Figure 6.4, in which we shall take the field amplitude to increase towards the left.

When the force is towards the right the particle is to the left of its central position, and when the force is to the left it is to the right of this position. Since the field amplitude increases towards the left the force on the particle is greater during the half cycle when it is acting towards the right and on average there is a force towards the right.

When the field gradient is perpendicular to the field direction the

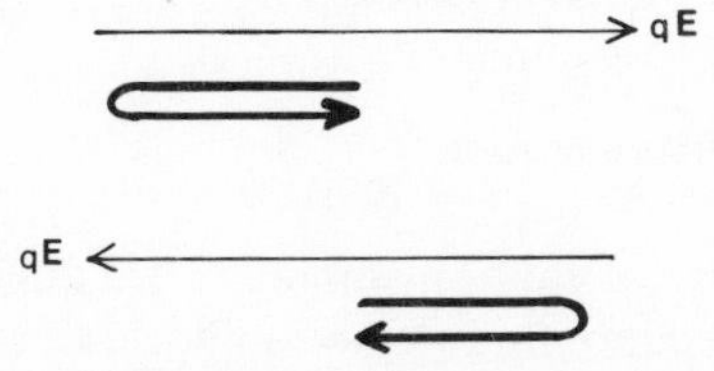

Figure 6.4 Particle orbit during successive half cycles of the oscillating field.

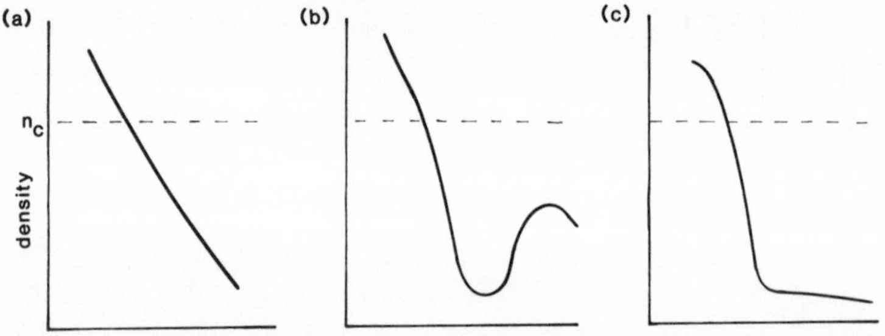

Figure 6.5 Creation of steep density gradient near critical surface: (*a*), initial density gradient; (*b*), cavity produced near critical surface; (*c*), final steep density gradient.

ponderomotive force arises from the second term in (6.25), and comes from a non-zero average $\boldsymbol{v} \times \boldsymbol{B}$ force. In a general field geometry both mechanisms will contribute, but regardless of which is important the ponderomotive force is given by the same expression (6.26). The force acts predominantly on the electrons, but if electrons are moved in the plasma an electric field is produced to drag the ions along and maintain neutrality. The ponderomotive force thus behaves like a pressure acting to force the plasma out of high field regions.

In laser–plasma interactions the process of resonant absorption at the critical surface, discussed in section 4.8, produces high-amplitude plasma waves in this region. These expel plasma from the region of the critical surface, and computer studies of the behaviour show a series of events as illustrated in Figure 6.5.

A cavity is initially created at the critical surface, but as the plasma flows outwards through the critical surface, the plasma on the low density side is swept away to leave a profile like that of Figure 6.5(*c*). The pressure jump resulting from the jump in density is balanced by the ponderomotive force of the electromagnetic field on the low-density side of the critical surface. Density jumps of this type have been observed in experiments, with the density going from half critical to above critical in a length of the order of the vacuum wavelength of the radiation.

This steep jump has some important consequences. One of these is that it increases the effectiveness of resonant absorption, making it important over a wide range of angles. Another is that the parametric decay of the incident wave into a plasma wave and an ion sound wave, which is possible only close to the critical surface, tends to be suppressed.

6.6 Modulation and filamentation

An electromagnetic wave of uniform amplitude in a uniform plasma is subject to instabilities which tend to produce non-uniformities. Modulation refers to an amplitude variation along the direction of propagation, producing the

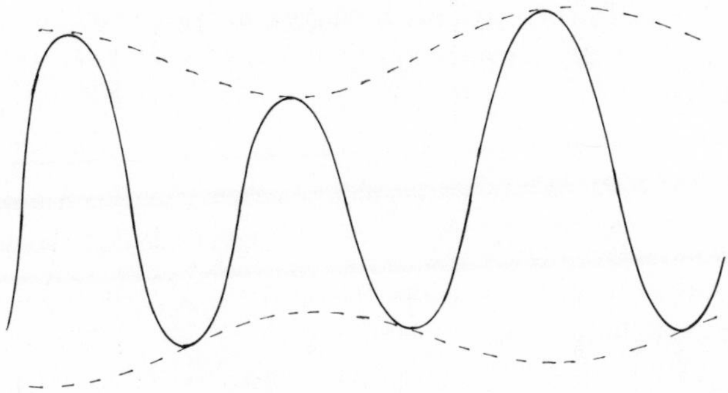

Figure 6.6 Modulation of wave.

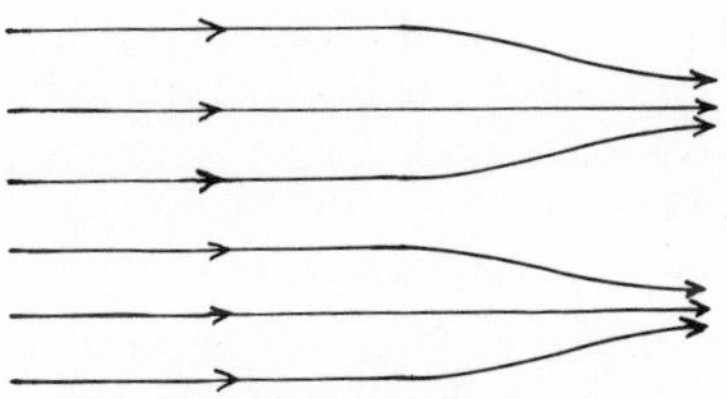

Figure 6.7 Filamentation

effect illustrated in Figure 6.6, while filamentation refers to a break up of the wave transverse to its direction of propagation, so that an initially uniform wavefront is channelled into narrow filaments as shown in Figure 6.7.

Both effects are simply explained in terms of the ponderomotive force and are a consequence of the fact that the refractive index of a plasma varies with the density in such a way as to focus electromagnetic waves into regions of low density. Any local increase in the wave intensity produces a low-density region, as a result of the ponderomotive force pushing plasma towards regions of weaker field. This, in turn, enhances the wave intensity so that the initial fluctuation in wave amplitude grows.

To consider these phenomena in a little more detail we consider a light wave in a plasma, the electric field of which obeys the equation

$$\frac{\partial^2 E}{\partial t^2} - c^2\frac{\partial^2 E}{\partial x^2} + \omega_p^2 E = 0 \qquad (6.27)$$

if the wave is taken to be travelling in the x-direction. In order to consider modulation of a plane wave solution, it is natural to take

$$E = A(x, t)\,e^{ikx - i\omega t} \qquad (6.28)$$

with ω and k related through the usual dispersion relation $\omega^2 = \omega_p^2 + k^2c^2$. If we consider perturbations about a uniform amplitude A_0 and if the plasma is assumed isothermal and the amplitude changes are slow enough that a steady-state pressure balance is obtained, then the change in density δn produced by the ponderomotive force is given by

$$\delta n \kappa T = -\frac{1}{4}\frac{e^2}{m\omega^2}(|A|^2 - |A_0|^2). \tag{6.29}$$

This simply expresses the balance of the pressure perturbations against the change in radiation pressure. Thus, if ω_{p0}^2 is the plasma frequency in the unperturbed system,

$$\omega_p^2 = \omega_{p0}^2\left(1 + \frac{\delta n}{n_0}\right)$$
$$= \omega_{p0}^2(1 + q(|E|^2 - |E_0|^2)),$$

where

$$q = -\frac{1}{4}\frac{e^2}{m\omega^2 n_0 \kappa T}.$$

Now, substituting (6.28) into (6.27) and neglecting second derivatives with respect to time gives

$$-2i\omega\frac{\partial A}{\partial t} - 2ic^2k\frac{\partial A}{\partial x} - c^2\frac{\partial^2 A}{\partial x^2} + \omega_{p0}^2 q(|A|^2 - |A_0|^2)A = 0.$$

Changing to a frame of reference moving with the group velocity of the wave, which is $c^2\omega/k$, eliminates the second term in this equation and leaves us with

$$i\frac{\partial A}{\partial t} + p_0\frac{\partial^2 A}{\partial \xi^2} + q_0(|A|^2 - |A_0|^2)A, \tag{6.30}$$

where

$$p_0 = \frac{c^2}{2\omega}, \quad q_0 = -\frac{\omega_{p0}^2 q}{2\omega}, \quad \xi = x - \frac{c^2\omega}{k}t.$$

An equation of this form is known as a nonlinear Schrödinger equation, on account of its similarity to the usual Schrödinger equation of quantum mechanics. A great deal of work has been done on analysing its properties, only a few of which we shall consider here. First let us examine the stability of the plane wave solution $A = A_0$ by linearizing about it. If we let

$$A - A_0 = a_1 e^{iK\xi - i\Omega t} + a_2 e^{-iK\xi + i\Omega t}$$

then the component of the linearized form of (6.30) going as $e^{iK\xi - i\Omega t}$ gives

$$\Omega a_1 - p_0K^2 a_1 + q_0|A_0|^2 a_1 + q_0 a_2 A_0^2 = 0$$

while that going as $e^{-iK\xi + i\Omega t}$ gives

$$-\Omega a_2 - p_0K^2 a_2 + q_0|A_0|^2 a_2 + q_0 a_1 A_0^{*2} = 0.$$

Combining these we obtain the dispersion relation

$$\Omega^2 = p_0^2K^4 - 2p_0q_0|A_0|^2K^2. \tag{6.31}$$

A growing solution is obtained if $p_0 q_0 > 0$, which is the case for the system under consideration, provided that K is small enough that

$$0 < K^2 < 2q_0 |A_0|^2 / p_0. \tag{6.32}$$

Recalling that A is the amplitude of a wave going as $e^{ikx - i\omega t}$ we see that the growth of a sinusoidal perturbation in A represents growth of a modulation in the wave amplitude after the fashion of Figure 6.6. From (6.32) it follows that the wavelength of the modulation may be expected to decrease as the wave amplitude increases. The filamentation instability may be analysed similarly using the three-dimensional analogue of (6.30) with a perturbation transverse to the propagation direction.

The nonlinear Schrödinger equation (6.30) is one of a number of nonlinear partial differential equations which have received a great deal of attention and which share a number of interesting properties. One such property is the existence of exact nonlinear soliton solutions, having the form of a solitary wave. It may be verified that an exact solution of (6.30) is

$$A(\xi, t) = \left(\frac{2\nu}{q_0}\right)^{1/2} \operatorname{sech}\left[\left(\frac{\nu}{p_0}\right)^{1/2} (\xi - \xi_0 - Ut)\right] \times \exp\left[i\left(\nu t + \frac{U}{2p_0}\xi - \frac{U^2}{4p_0}t + \theta_0\right)\right]$$

with ν, ξ_0, U and θ_0 independent parameters. The amplitude, determined by the sech factor, is of the shape of a solitary wave travelling with speed U. The solution here is an envelope soliton, the solitary wave shape being that of the envelope of the oscillations. For other systems described by different equations the solution may be the actual physical shape of the amplitude.

If the initial state of the system consists of two such solitons, far apart so that they do not overlap and having velocities which cause them to come together, then they pass through each other and emerge with their initial shapes intact, the interaction only causing an overall displacement in position of the solitons. Since the system is nonlinear and does not obey the principle of superposition familiar in linear systems, this is rather remarkable behaviour. The class of equations to which the nonlinear Schrödinger equation belongs can be solved by a technique known as the method of inverse scattering, which sheds light on this property of solitons and other features of the solutions of these equations. Further discussion of this method is, unfortunately, beyond the scope of this work.

6.7 Computational methods—fluid codes

Many problems in plasma physics involve nonlinear behaviour in complicated geometries, making analytic progress almost impossible. Under these circumstances the use of numerical techniques to analyse the plasma behaviour becomes essential, and much effort has been devoted to the writing of computer codes for the solution of various plasma physics problems. In this section we shall discuss some of the essential ideas involved in applying numerical methods to a fluid description of the plasma. Most work on plasma physics has made use of finite difference techniques for the numerical solution of the fluid equations, and we shall confine our attention to these, although other methods, for example finite element techniques, are available for the solution of partial differential equations. Even with this restriction, a comprehensive survey of numerical methods for the solution of the fluid equations could well be the subject of a book in itself, and in this brief discussion we merely try to give some idea of what is involved and of some of the difficulties to be overcome before a successful code is obtained.

The equations of the fluid contain only first-order derivatives with respect to time, so the system of equations is of the form

$$\frac{\partial u}{\partial t} = F(u), \tag{6.33}$$

where u is a vector containing the various independent variables describing the system—density, velocity etc. In an Eulerian difference scheme we calculate the values of the independent variables at a set of points fixed in space. Taking, for simplicity a one-dimensional system, these would generally be equally spaced throughout the domain of interest. Then we aim to find the values at each of these points at equally spaced points in time. In the (x, t) domain we are thus interested in the values at a series of points on a rectangular mesh as illustrated in Figure 6.8. The index i will be used for the spatial coordinate and n for the time, as shown. At the initial time $t = 0$ the variables are specified, and we then need some way of stepping forward to Δt, $2\Delta t$ and so on. To do this we

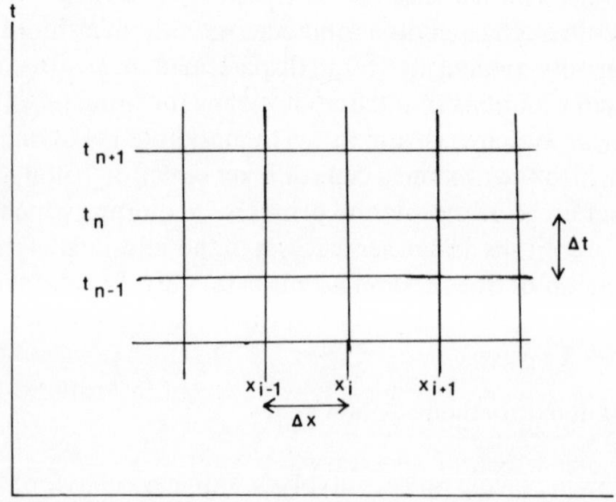

Figure 6.8 Mesh for numerical solution of one-dimensional system.

let u_i^n be the value of u at (x_i, t_n) and approximate $\partial u_i^n/\partial t$ in (6.33) by

$$\frac{u_i^{n+1} - u_i^n}{\Delta t} \tag{6.34}$$

In general $F(u)$ will contain first or second derivatives with respect to x, and we approximate these in terms of the values at the mesh points. For example we might try

$$\frac{\partial u_i^n}{\partial x} \approx \frac{u_{i+1}^n - u_{i-1}^n}{2\Delta x} \tag{6.35}$$

$$\frac{\partial^2 u_i^n}{\partial x^2} \approx \frac{u_{i+1}^n - 2u_i^n + u_{i-1}^n}{(\Delta x)^2} \tag{6.36}$$

If we use the Taylor expansions

$$u_{i+1}^n = u_i^n + \frac{\partial u_i^n}{\partial x}\Delta x + \frac{1}{2}\frac{\partial^2 u_i^n}{\partial x^2}(\Delta x)^2 + \cdots$$

$$u_{i-1}^n = u_i^n - \frac{\partial u_i^n}{\partial x}\Delta x + \frac{1}{2}\frac{\partial^2 u_i^n}{\partial x^2}(\Delta x)^2 - \cdots,$$

then we can see that (6.35) and (6.36) are correct to order $(\Delta x)^2$. In these equations we have used the values of u at the nth time step to evaluate the spatial derivatives. This defines what is known as an explicit scheme, since by using the approximations (6.34)–(6.36) an explicit formula for the values at time t_{n+1} in terms of those at time t_n is obtained, and so we can step forward in time easily. Suitable boundary conditions must be imposed at the boundaries of the x-domain, for example u might be fixed at these points.

An explicit scheme has the advantage of simplicity, but is subject to restrictions on the allowed time step. To see why this is so we recall that the fluid equations are of a type which permit the propagation of disturbances with characteristic wave speeds. If c_0 is the fastest of these speeds, then we may expect the value of u at (x, t_{n+1}) to depend on the values at t_n in the range $x_i - c_0\Delta t \leqslant x < x_i + c_0\Delta t$, assuming for the moment that the fluid is at rest. Effects at points further from x_i cannot propagate to x_i within the time Δt. Now, at the time t_n we are using the points x_{i+1}, x_i and x_{i-1} to calculate the spatial derivatives. If the value of Δx is less than $c_0\Delta t$, then we are forcing the value of u at (x_i, t_{n+1}) to depend on its values at t_n over a smaller spatial range, namely $2\Delta x$, than it should in principle, and so we cannot expect to obtain correct results. Thus, we may expect to have to impose the condition $\Delta x \geqslant c_0\Delta t$, or $\Delta t \leqslant \Delta x/c_0$ in order to obtain acceptable results. A more careful treatment does in fact yield this condition, the Courant–Friedrichs–Lewy condition, as one which is necessary for stability of an explicit numerical scheme. If it is not satisfied errors in the solution grow exponentially and rapidly dominate the behaviour. In a fluid with flow velocity v, it becomes, for fairly obvious reasons, $\Delta t \leqslant \Delta x/(|v| + c)$. If one has decided how many spatial grid points are necessary to give a reasonably accurate representation of the solution then the Courant–Friedrichs–Lewy condition imposes an upper limit on the time step used which, in turn, imposes a limit on the time for which plasma behaviour can be followed within the available computer time.

Further difficulties associated with this type of numerical solution arise in the treatment of the convective derivatives which are characteristic of the fluid equations. To illustrate some of these let us consider the simple equation

$$\frac{\partial u_i}{\partial t} + a\frac{\partial u}{\partial x} = 0, \tag{6.37}$$

where u is now a single variable and a is a constant. The solution of this equation is

$$u(x, t) = u(x - at, 0),$$

so that it describes convection of any initial disturbance with a constant speed a.

Now, let us replace it with its finite difference approximation

$$\frac{u_i^{n+1} - u_i^n}{\Delta t} = -a\frac{u_{i+1}^n - u_{i-1}^n}{2\Delta x}, \tag{6.38}$$

and consider a solution of the form $u_0\,e^{ikx - i\omega t}$. Substituting this into (6.38) gives

$$\frac{e^{-i\omega\Delta t} - 1}{\Delta t} = -a\frac{e^{ik\Delta x} - e^{-ik\Delta x}}{2\Delta x}$$

or

$$e^{-i\omega\Delta t} = 1 - \frac{ia\Delta t}{\Delta x}\sin(k\Delta x).$$

Now, $e^{-i\omega\Delta t}$ is the factor by which the solution is multiplied at each time step, and since $|e^{-i\omega\Delta t}|^2 = 1 + a^2(\Delta t/\Delta x)^2 \sin^2(k\Delta x)$, which is greater than one for almost all values of k, the amplitude of the solution grows in time, regardless of the value of Δt. This numerical scheme is therefore unstable, a problem which can be got round by replacing the centre difference scheme on the right-hand side of (6.38) by an upstream difference, that is, taking a to be positive, we use the approximation

$$\frac{\partial u_i^n}{\partial x} \approx \frac{u_i^n - u_{i-1}^n}{\Delta x} \tag{6.39}$$

using only the value on the upstream side of the ith point. If we look again at the behaviour of a plane wave solution we get

$$\frac{e^{-i\omega\Delta t}}{\Delta t} = -a\frac{1 - e^{-ik\Delta x}}{\Delta x},$$

or

$$e^{-i\omega\Delta t} = 1 - 2ia\frac{\Delta t}{\Delta x}e^{-ik\Delta x/2}\sin\left(\frac{k\Delta x}{2}\right). \tag{6.40}$$

This time

$$|e^{-i\omega\Delta t}|^2 = 1 - 4\lambda(1-\lambda)\sin^2(k\Delta x/2),$$

with $\lambda = a\Delta t/\Delta x$. If $\lambda \leqslant 1$ this is less than or equal to unity and so the scheme is stable. The condition $\lambda \leqslant 1$ is just the Courant–Friedrichs–Lewy condition, a being the velocity of propagation of a disturbance in the system.

Stability of the system does not, however, guarantee that it gives an accurate solution of the problem. The exact solution is just $u_0\,e^{ikx-i\omega t}$, with $\omega = ka$, and to reproduce this we ought to have all Fourier components propagating at the same speed with constant amplitude. Instead the amplitude of each Fourier component is multiplied between successive time steps by a complex number whose amplitude is generally less than one. Thus, an initial disturbance decays. The decay is faster for larger k, so that the initial effect is a smoothing out of short wavelength disturbances. This is typical of a diffusion process and represents a numerical diffusion introduced into the system by the finite differencing. Also the ratio of the real part of ω to k calculated from (6.40) is not constant, so that dispersion has also been introduced into the system. The origin of the diffusion and dispersion can be seen if we note that

$$\frac{u_i^n - u_{i-1}^n}{\Delta x} = \frac{\partial u_i^n}{\partial x} - \frac{\Delta x}{2}\frac{\partial^2 u_i^n}{\partial x^2} + \frac{(\Delta x)^2}{6}\frac{\partial^3 u_i^n}{\partial x^3} \ldots,$$

so that we are actually solving, instead of (6.37), an equation with higher derivatives appearing. The second derivative is characteristic of a diffusion equation and the third derivative contributes to dispersion.

A great deal of effort has been expended in devising more complicated difference schemes which reduce the effects of numerical diffusion and dispersion and give an acceptable approximation to the true solution. One problem with the Eulerian schemes described above is the tendency of initially smooth gradients in a fluid to steepen and form shock waves. Once the gradient is on a scale comparable to the distance between adjacent points of the mesh the solution cannot be represented accurately. Special care is needed to maintain accuracy away from the shock front, even if detail within the shock cannot be followed.

An alternative approach which is useful when large gradients develop is the Lagrangian scheme, in which the spatial mesh points are not fixed, but move with the fluid. If an initially smooth gradient steepens, the mesh points in the region of steepening become more closely spaced. Lagrangian schemes are quite easy to implement in one dimension, with each spatial mesh point moving according to the equation

$$\frac{dx}{dt} = v(x, t),$$

and convective derivatives becoming total derivatives moving with the mesh point. Their generalization to more than one dimension is, however, more difficult than with Eulerian schemes. In complicated flows an initially rectangular mesh becomes more and more convoluted and periodic re-zoning back to a more regular mesh may be required. This, of course, involves some error in interpolating from the original mesh to the new mesh.

The Courant–Friedrichs–Lewy condition may impose an unacceptably small time step on explicit schemes. In effect it means that the time step must be small enough to resolve the propagation of the fastest-moving wave across the system. This is fine if the object is to study such high-frequency effects, but if we want to study phenomena occurring on a longer time scale, without knowing all the details of fast oscillations, it may require a time step so small as to make the calculation impossible. The answer under these circumstances is to use an implicit scheme, in which the spatial derivatives are calculated not at time t_n but at time t_{n+1}. These involve the unknown values u_i^{n+1} and so the finite difference version of (6.33) contains unknowns on both sides. The result is a set of equations for the u_i^{n+1} which must be solved by inversion of a matrix. In one dimension and with approximations like (6.35) and (6.36) which involve only u_{i+1}^{n+1} and u_{i-1}^{n+1}, the matrix is tridiagonal, that is its only non-zero elements lie on the main diagonal and the two immediately adjacent diagonals. There is a simple algorithm for inverting such matrices. In more than one dimension the matrix is not so simple, however, and various more complicated techniques have to be used.

As well as being stable and reasonably accurate it is important that a numerical scheme for solving the fluid equations, especially in more than one dimension, be as efficient as possible, in the sense of requiring as few as possible computer operations at each time step. Even with the largest and most modern computers these codes are very time-consuming and expensive to run. On the other hand there are many problems of the greatest importance involving the nonlinear behaviour of plasmas which are quite intractable by any other technique.

6.8 Computational methods—particle codes

As was discussed in the previous chapter, small-scale, high-frequency oscillations and instabilities in a plasma require a description in terms of the distribution of particles in a velocity space rather than in terms of fluid quantities. To investigate the nonlinear behaviour of these effects requires appropriate numerical techniques. One approach is to solve the kinetic equations describing the system, for example the Vlasov equation, by numerical techniques similar to those used for the fluid equations, but a more common approach has been by way of particle simulations.

The basic idea is to take a group of particles and to follow their motions in

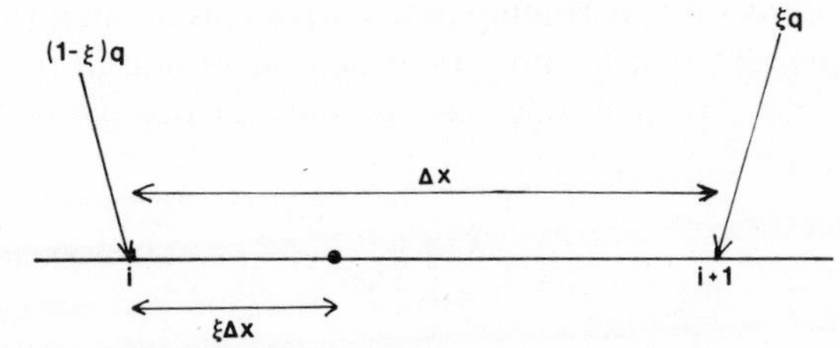

Figure 6.9 Division of charge between neighbouring mesh points in a one-dimensional PIC code.

the self-consistent electric and magnetic fields which they produce. However, if we calculate the force on each particle due to the other particles, then we have to look at $N(N-1)/2$ pairwise interactions in a system of N particles. Such calculations are done to investigate topics like the statistical mechanics of liquids and dense plasmas, but the computational time involved goes as N^2 and they are restricted to a few hundred particles. For plasmas, in which the object is usually to investigate collective effects on a scale longer than the Debye length and involving large numbers of particles, this approach is impracticable.

The answer to this difficulty is to use what are called particle-in-cell (PIC) codes. In these a spatial grid is defined, just as described in section 6.6 and the charge of a particle in a cell of this grid is divided up amongst its neighbouring grid points. This is illustrated in Figure 6.9 for a one-dimensional system.

The fields are then calculated by finite-differencing of Poisson's equation, in an electrostatic code, or of the full set of Maxwell's equations. This yields the fields at the grid points. The fields at the position of each particle are then calculated by linear interpolation and the position and velocity of each particle updated to the next time step using Newton's law. With this procedure the number of operations per time step becomes proportional to N for a fixed number of grid points. There should be a large number of particles per cell, so that the fields do not have sudden discontinuities as a particle crosses a cell boundary, and a typical code will use around 10^5 or 10^6 particles.

The procedure adopted in PIC codes amounts to a smoothing of the fields, precisely what is required in order to simulate a collisionless plasma with a large number of particles per Debye sphere. The numerical techniques involved are, in principle at least, quite simple and do not suffer unduly from numerical instability. One interesting instability is associated with the existence of the spatial grid. In one dimension we note that a disturbance of wavenumber k cannot be distinguished from those with wavenumber $k + 2\pi n/\Delta x$ where n is an integer and Δx the grid spacing. These are known as aliases of the original wave. The effect of the redistribution of charge to the grid points is to couple together these different modes, leading to normal modes of the plasma which are not plane waves but linear combinations of the aliases. In some circumstances these modes may grow in time, giving an aliasing instability. This problem may be avoided by ensuring that the aliases have phase velocities less than the thermal velocity of the particle and so are strongly damped. This requires

$$\left(k+\frac{2\pi n}{\Delta x}\right)\lambda_D \rightleftharpoons 1$$

where λ_D is the Debye length. Assuming that the modes we wish to examine have $k\lambda_D \ll 1$ and so are not heavily damped, we therefore wish to have $2\pi\lambda_D/\Delta x \gtrsim 1$. This puts a restriction on the size of the spatial grid, which should not contain too many Debye lengths.

Limits on computer time and storage generally mean that plasma simulations can only look at small plasma regions. They are however very useful in elucidating the nonlinear behaviour of small-scale phenomena like microinstabilities and high-frequency waves. One field in which they have been very fruitful is the study of laser–plasma interactions where very high-intensity wave fields are present. An an example of the sort of phenomenon which may be investigated we consider the behaviour of the large-amplitude plasma waves which are excited near the critical surface by resonant absorption (discussed in section 4.8). In such a plasma wave electrons with substantial velocities relative to the wave phase velocity may be trapped in the potential troughs. If this trapping region extends into the main part of the electron distribution, as shown in Figure 6.10, then particles are taken from the bulk of the distribution and may have velocities extending up to the upper limit of the trapping region when they are untrapped by collisions or by reaching the edge of the region in which the plasma wave is excited. The effect of resonant absorption is thus to generate a high-energy tail on the electron distribution function. Such behaviour is easy to describe qualitatively, but is very difficult to obtain any convincing analytic theory which describes the process in detail and gives, for example, predictions of the exact form of the

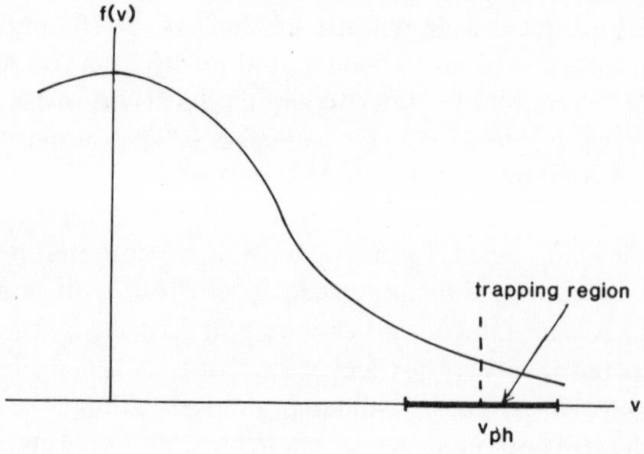

Figure 6.10 Trapping of electrons by a large-amplitude plasma wave.

electron distribution function. This is just the sort of problem for which a PIC code comes into its own.

6.9 Concluding remarks

This chapter has touched on some of the important techniques used to analyse nonlinear behaviour, but is in no way an exhaustive survey. Attempts to understand the nonlinear behaviour of instabilities play a key role in the investigation of plasmas. While linear theory can give an indication of the instabilities which may occur it is the level at which they saturate nonlinearly and the effect of the saturated oscillations on the plasma which are important in practice.

One of the central problems in tokamak behaviour is the fact that transport of energy across the confining magnetic field exceeds, by up to two orders of magnitude, that which is calculated to result from particle collisions. This and other so-called anomalous effects are thought to be due to the existence of fluctuations in the plasma and are the subject of a great deal of research. It is easy to see intuitively how field fluctuations could scatter particles and produce enhanced diffusion, but a quantitative calculation poses a very difficult problem.

An interesting feature of nonlinear equations in general, and one which is presently attracting a great deal of interest, is their tendency to exhibit chaotic behaviour where well-defined deterministic equations give rise to solutions which have the appearance of random fluctuations with a broad noise-like spectrum. In a regime exhibiting such solutions small changes in initial conditions produce wide variations in the solutions at later times. It is possible that the apparently random behaviour seen in turbulent fluids and plasmas may be associated with this phenomenon.

Problems

6.1 Show that a resonant triad of collinearly propagating waves is possible if when (ω, k) curves are drawn for the modes involved it is possible to draw a parallelogram connecting the three (ω, k) values involved and the origin.

6.2 Illustrate the above construction for Raman and Brillouin scattering, remembering that the backward propagating wave should be taken to have negative k.

6.3 Compute some numerical solutions of (6.16) and the corresponding equations with damping included. Try to explain the qualitative features of the solutions you obtain.

6.4 Write computer programs to solve the equations (6.37) with step function initial conditions, using the finite difference schemes of (6.38) and (6.39). Note the difference between the stable and unstable algorithms and the effects of numerical diffusion and dispersion on the stable scheme.

7 Diagnostic techniques

7.1 Introduction

In most of this book we have been concerned with a discussion of the basic properties of a plasma and have referred to experimental results only in passing, without giving details of how they are obtained. Here we redress the balance somewhat by discussing diagnostic techniques, that is the methods which are used to obtain information about actual plasmas. Since many plasmas of interest are under rather extreme conditions of temperature and are often of very short duration, it is by no means a simple matter to obtain details of the conditions which prevail in them.

Various diagnostic methods are available and may be divided into the following main classes. The first (which will be discussed in section 7.2) is the use of probes, i.e. electrodes and coils of various sorts inserted into the plasma or placed in its neighbourhood. Probes placed in the plasma are most useful for low-temperature dense plasmas, in which the presence of a probe does not produce too great a change in the ambient conditions. Next are methods involving electromagnetic radiation. Information may be obtained simply by observing tne radiation emitted as discussed in section 7.3, or by looking at the scattering of radiation sent into the plasma from a laser or other source, as described in section 7.4. Finally there are measurements on particles emitted from the plasma, discussed in section 7.5.

We shall give the basic principles of some of the most important diagnostic techniques in these various classes without going into the technical details of the experimental methods. It will be apparent that some of the most important parameters, like the temperature and density, can be measured in more than one way. Large experiments make use of diagnostics of all the types discussed here, so that there is the possibility of cross-checking of data obtained in different ways.

7.2 Probes

We begin by considering the use of electrostatic or Langmuir probes, one of the longest-established methods of investigating a plasma. These are simply small electrodes inserted into the plasma. In the case of a plasma produced in a discharge tube, one of the electrodes producing the discharge can be used as a

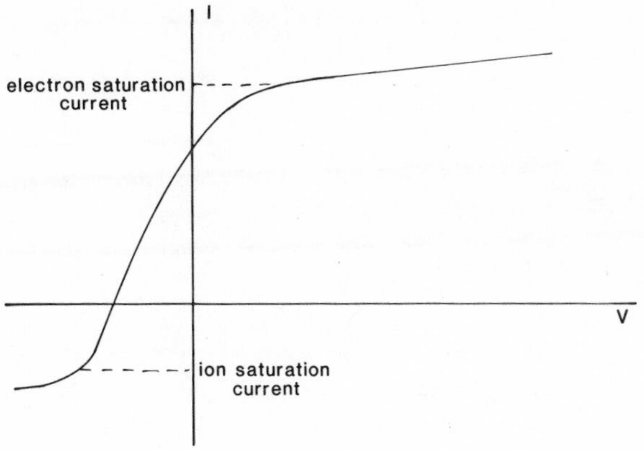

Figure 7.1 Current–voltage characteristic of a Langmuir probe.

reference level which can be related to the potential of the plasma. The current flowing between the probe and the reference electrode is then measured as a function of the potential of the probe with respect to the electrode. A typical curve takes the shape shown in Figure 7.1.

The behaviour of the probe may be explained as follows. When the electrode has a high positive potential with respect to the ambient plasma, electrons are attracted to it, and form a sheath of negative charge around it, extending out to a distance of the order of a Debye length from the probe. Beyond this distance there is essentially no field due to the probe and the current is limited by the flux of electrons arriving at the boundary of the sheath as a result of their random thermal motion. This current, almost independent of the potential, is the electron saturation current. This is proportional to the product of the density and the average electron speed, and so to $n_e T_e^{1/2}$.

If the probe potential is decreased below the ambient potential, only electrons with sufficient kinetic energy can reach it and the current decreases. The slope of the current–voltage characteristic in this region, which corresponds to the steep part of the curve in Figure 7.1, is related to the slope of the electron distribution as a function of energy, and so gives the electron temperature. The potential at which the current goes to zero is called the floating potential. It is below the ambient potential, since otherwise electrons arrive at the surface of the probe at a greater rate than ions, owing to their greater thermal velocity. When the potential becomes sufficiently negative that a negligible number of electrons can reach it, a sheath of positive charge is set up around it and the current levels off. The value of the current at which this occurs is the ion saturation current. Its value depends on the potential required to repel the bulk of the electrons, i.e. on the electron temperature rather than the ion temperature. From the shape of the characteristic it is thus possible to ascertain both the plasma temperature and density.

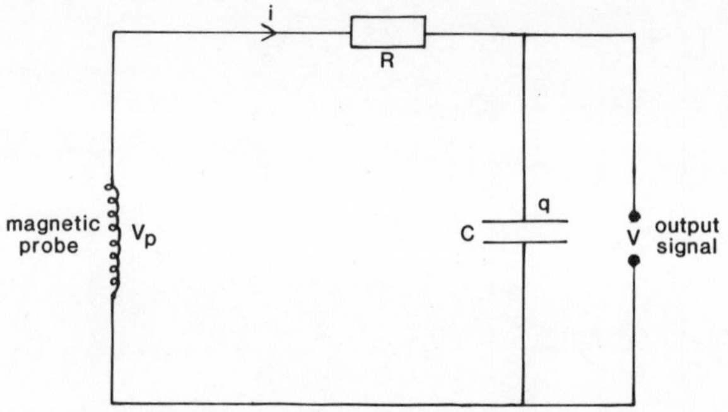

Figure 7.2 Integration network for magnetic probe.

In a plasma geometry where there is no suitable reference electrode, we may use a double probe, to which similar principles apply.

The local magnetic field in or around a plasma may be measured using a small solenoid. If this has N turns each of area a, then the voltage which it generates is given by

$$V_p = Na\frac{\mathrm{d}B}{\mathrm{d}t},$$

where B is the magnetic field component along its axis. An output signal proportional to B rather than to its rate of change can be obtained by using a simple integration network as shown in Figure 7.2.

If q is the charge on the capacitor, and the induction of the coil is negligible, then

$$V_p = iR + \frac{q}{C} = R\frac{\mathrm{d}q}{\mathrm{d}t} + \frac{q}{C}$$

and so, since $V = q/C$,

$$\frac{\mathrm{d}}{\mathrm{d}t}(V\mathrm{e}^{t/RC}) = \frac{V_p}{RC}\mathrm{e}^{t/RC} = \frac{Na}{RC}\frac{\mathrm{d}B}{\mathrm{d}t}\mathrm{e}^{t/RC}.$$

Thus

$$V = \frac{Na}{RC}\mathrm{e}^{-t/RC}\int_0^t \frac{\mathrm{d}B(t')}{\mathrm{d}t'}\mathrm{e}^{t'/RC}\,\mathrm{d}t',$$

and if we assume that $B = 0$ at $t = 0$ and integrate by parts we obtain

$$V = \frac{Na}{RC}B - \frac{Na}{(RC)^2}\mathrm{e}^{-t/RC}\int_0^t B(t')\mathrm{e}^{t'/RC}\,\mathrm{d}t'.$$

Choosing the time constant RC of the circuit to be much longer than the time scale over which B changes makes the second term small and we can

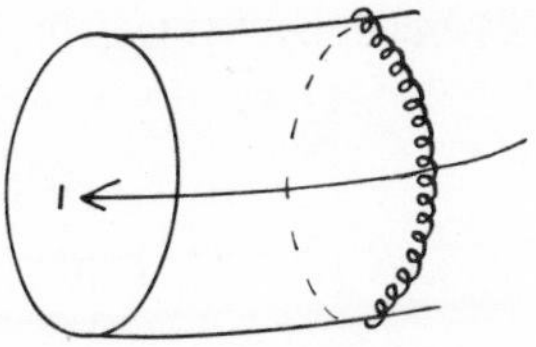

Figure 7.3 Rogowski coil.

therefore obtain an output given by

$$V = \frac{Na}{RC} B.$$

In a thermonuclear plasma probes inside the plasma perturb it too much by cooling it and introducing impurities. However, similar measurements made outside the plasma may yield important information. The total magnetic flux through any surface can be obtained from the voltage induced in a flux loop around its boundary, while the current flowing in a plasma can be measured by means of a Rogowski coil. The principle of this latter device is based on the fact that the total current in the plasma can be expressed as a line integral of the magnetic field around it as follows:

$$I = \int \boldsymbol{J} \cdot \mathrm{d}\boldsymbol{S} = \oint \nabla \times \boldsymbol{H} \cdot \mathrm{d}\boldsymbol{S} = \frac{1}{\mu_0} \oint \boldsymbol{B} \cdot \mathrm{d}\boldsymbol{l}.$$

The Rogowski coil is simply a solenoid wound around the plasma as shown in Figure 7.3. The voltage induced in it is proportional to the rate of change of the line integral of the magnetic field around the plasma, and so to $\mathrm{d}I/\mathrm{d}t$. An integration network as described above may be used to obtain an output proportional to the current. Information on the position and shape of the plasma in a toroidal system may be inferred from sufficiently detailed measurements of the magnetic field outside the plasma.

7.3 Radiation emission

One of the most powerful diagnostic techniques for plasmas is observation of the radiation emitted by them. This can be applied both to laboratory plasmas and to otherwise inaccessible plasmas like those on the Sun. Various emission mechanisms produce radiation, and a wide range of information can be obtained. We shall give brief discussions of the principles of bremsstrahlung, cyclotron emission and line radiation and indicate some of the information which may be obtained from their measurement.

Bremsstrahlung is the radiation emitted by electrons as they are accelerated during collisions with ions. Its detailed analysis requires a quantum mechanical calculation, but some of its principal features can be seen from quite simple classical ideas. The power radiated by an electron with charge e and velocity $\boldsymbol{v}$ is given by

$$P = \frac{e^2}{6\pi\varepsilon_0 c^3} \frac{\dot{v}^2 - (\boldsymbol{v} \times \dot{\boldsymbol{v}})^2/c^2}{(1 - v^2/c^2)^3} \tag{7.1}$$

or in the non-relativistic limit when $v/c \ll 1$,

$$P = \frac{e^2}{6\pi\varepsilon_0 c^3} \dot{v}^2. \tag{7.2}$$

If we consider an electron with speed v colliding with an ion whose velocity is negligible, the impact parameter of the collision being b, the maximum acceleration is of the order

$$a = \frac{Ze^2}{4\pi\varepsilon_0 b^2 m_e}$$

while the angle through which the electron is scattered is given (see section 3.6) by

$$\cot\frac{\psi}{2} = \frac{4\pi\varepsilon_0 m_e b}{Ze^2}$$

The total change in velocity during the collision is

$$\Delta v = \sqrt{2}v(1 - \cos\psi)^{1/2} = 2v\sin(\psi/2).$$

Defining an effective collision time by

$$\tau = \frac{\Delta v}{a},$$

and using the approximation $\sin(\psi/2) \approx 1/[\cot(\psi/2)]$ we obtain

$$\tau \approx 2b/v.$$

Thus the collision will give a pulse of radiation lasting a time τ, which if it is Fourier analysed in time will give its dominant contribution to that part of the frequency spectrum with

$$\nu \simeq \frac{1}{2\pi\tau}. \tag{7.3}$$

The energy radiated during the collision is

$$\Delta E \approx \frac{e^2}{6\pi\varepsilon_0 c^3} a^2\tau = \frac{Z^2 e^2}{48\pi^3\varepsilon_0^3 c^3 m_e^2 b^3 v}. \tag{7.4}$$

The average energy radiated per unit time by an electron is

$$\frac{\mathrm{d}E}{\mathrm{d}t} = \int_{b_{\min}}^{b_{\max}} \Delta E\, 2\pi n_i v b\,\mathrm{d}b = \frac{Z^2 e^6 n_i}{24\pi^2\varepsilon_0^3 c^3 m_e^2} \int_{b_{\min}}^{b_{\max}} \frac{\mathrm{d}b}{b^2}. \tag{7.5}$$

In earlier integrals over the impact parameter the upper limit was taken to be the Debye length. However in (7.5) there is no divergence at the upper end of the range, which may be taken to be at infinity. The lower limit appropriate to this calculation is the de Broglie wavelength, since for smaller impact parameters our classical picture of the electron orbit no longer holds. With $b_{\min} = \hbar/mv$ we obtain

$$\frac{\mathrm{d}E}{\mathrm{d}t} = \frac{Z^2 e^6 n_i}{24\pi^2 \varepsilon_0^3 c^3 m_e \hbar} \frac{v}{\hbar}.$$

The dominant contribution to (7.5) is from small values of b. Summing this over all electrons, averaged over a Maxwellian velocity distribution, we obtain the total bremsstrahlung loss per unit volume

$$W = \frac{Z^2 e^6 n_e n_i}{24\pi^2 \varepsilon_0^3 c^3 m_e \hbar} \left(\frac{m_e}{2\pi\kappa T}\right)^{3/2} 4\pi \int_0^\infty v^3 \, \mathrm{e}^{-m_e v^2/2\kappa T_e} \, \mathrm{d}v$$

$$= \frac{Z^2 e^6 n_e n_i}{12\pi^2 \varepsilon_0^3 c^3 m_e \hbar} \left(\frac{2\kappa T_e}{\pi m_e}\right)^{1/2}. \quad (7.6)$$

For diagnostic purposes the shape of the frequency spectrum is more important than the total emission. From (7.3) we see that the frequency of the emission is related to the impact parameter by

$$\nu = \frac{v}{4\pi b}$$

Since the minimum impact parameter is $\hbar/m_e v$, an electron of velocity v only contributes to those frequencies for which

$$\nu < \frac{m_e v^2}{4\pi\hbar} = \frac{\frac{1}{2} m_e v^2}{h}.$$

This can be interpreted as saying that the total energy of the electron must be greater than the energy of the emitted photon.

Now, radiation in the frequency range $\nu \to \nu + \mathrm{d}\nu$ is emitted by particles with impact parameters in the range $b \to b + \mathrm{d}b$ with

$$\mathrm{d}b = -\frac{v}{4\pi\nu^2} \mathrm{d}\nu.$$

so the number of the relevant collisions per unit time is

$$2\pi v b \, \mathrm{d}b = 2\pi v \frac{v}{4\pi\nu} \frac{v}{4\pi\nu^2} \quad \mathrm{d}\nu = \frac{v^3}{8\pi\nu^3} \mathrm{d}\nu.$$

Using the expression (7.4) we obtain the energy radiated per unit time in the above frequency range by a particle with velocity v as

$$\frac{\mathrm{d}E(\nu)}{\mathrm{d}t} = \frac{Z^2 e^6}{48\pi^3 \varepsilon_0^3 c^3 m_e^2}\left(\frac{v}{4\pi\nu}\right)^{-3}\frac{1}{v}\frac{v^3}{8\pi\nu^3}\,\mathrm{d}\nu$$

$$= \frac{Z^2 e^6}{6\pi\varepsilon_0^3 c^3 m_e^2}\frac{1}{v}\,\mathrm{d}\nu$$

The total power radiated in this frequency range is $W(\nu)\,\mathrm{d}\nu$, with

$$W(\nu) = \frac{Z^2 e^6}{6\pi\varepsilon_0^3 c^3 m_e^2}\left(\frac{m_e}{2\pi\kappa T_e}\right)^{3/2}\int_{(2h\nu/m_e)^{1/2}}^{\infty} v\mathrm{e}^{-m_e v^2/2\kappa T_e}\,\mathrm{d}v$$

$$= \frac{Z^2 e^6 n_e n_i}{6\pi\varepsilon_0^3 c^3 m_e^2}\left(\frac{m_e}{2\pi\kappa T_e}\right)^{1/2}\mathrm{e}^{-h\nu/\kappa T_e}. \tag{7.7}$$

The integral over frequency of $W(\nu)$ is just the total power given by (7.6). A more careful semi-classical calculation gives $1/\sqrt{3}$ times (7.7), while quantum calculations introduce a further factor $g(\nu, T_e)$ known as the Gaunt factor. For most parameters of interest the Gaunt factor is around one.

From (7.7) we see that the slope of a plot of log $W(\nu)$ against ν gives the electron temperature, while the density may be also inferred if the absolute intensity of the radiation is known. Deviations of the plot from a straight line indicate deviations from Maxwellian in the energy distribution of the particles. The wavelength range of interest for bremsstrahlung measurements ranges from the ultraviolet for comparatively cold plasmas into the X-ray region for thermonuclear plasmas.

The power radiated by bremsstrahlung is proportional to Z^2. For this reason a comparatively small concentration of heavy, highly ionized impurity atoms in a plasma may lead to a very large increase in radiation losses. One of the main problems to be faced in obtaining the conditions required for thermonuclear fusion in magnetic containment devices is the reduction of impurities to a sufficiently low level.

In a magnetic field charged particles produce synchrotron radiation as a result of the acceleration involved in their gyration about magnetic field lines. Electron cyclotron emission may provide a useful diagnostic in high-temperature plasmas. The spectrum of the emission consists of peaks centred around the harmonics of the cyclotron frequency, with broadening of the peaks being the result of the Doppler shift due to the parallel motion and the relativistic mass shift, as discussed for absorption in section 5.8. In a thermal equilibrium plasma, absorption and emission are related through Kirchoff's Law.

The emission around the cyclotron frequency or its harmonics depends on the plasma temperature. In a non-uniform magnetic field different frequencies come from different parts of the plasma and so there is the possibility of obtaining space-resolved temperature data from electron cyclotron emission measurements. We shall not pursue the details further, but simply note that such measurements are now a routine diagnostic on most large tokamaks.

Synchrotron radiation is also an important part of the emission observed from radio galaxies. The spectral characteristics and polarization of a large part of the observed radiation is not consistent with thermal emission, but can be explained in terms of synchrotron radiation from relativistic electrons. This has provided important evidence for the existence of large-scale magnetic fields in the universe.

Finally in this section on the radiation emitted by a plasma we should mention the line radiation coming from electron transitions between bound states. Such lines are, of course, characteristic of the ions and neutral atoms producing them and provide information about the composition of the plasma. Thus in fusion research spectroscopic measurements play an important role in investigating the concentration and diffusion of impurity species. The details of the line shapes can also provide information about the plasma. Ion motion, for instance, produces a Doppler shift of the frequency, so with an assembly of randomly moving ions there is a Doppler broadening of a spectral line from which, in principle, the ion temperature may be obtained. However other effects also give rise to line broadening, notably the Zeeman effect due to the magnetic field and Stark broadening produced by the microscopic electric fields in the plasma. Spectroscopy plays an important role in the investigation of dense laser-compressed plasmas, whose small size, high density and short duration makes many of the other diagnostic techniques difficult or impossible.

7.4 Radiation scattering and interferometry

In addition to passive observation of radiation spontaneously emitted from the plasma, use may be made of scattering of a beam of radiation sent into the plasma. Schematically the experimental set-up is as in Figure 7.4.

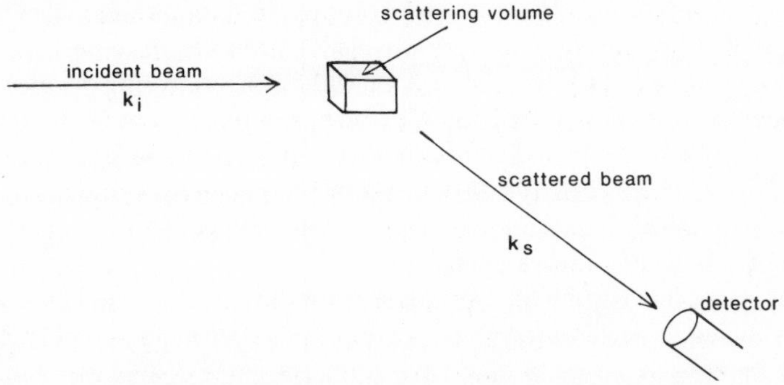

Figure 7.4 Schematic of radiation scattering experiment.

The mechanism producing scattering is quite straightforward. Electrons oscillate in the electric field of the incident wave and act as dipole radiators, emitting radiation in all directions. With a completely uniform distribution of scattering centres the net scattering is zero, since the various contributions cancel each other. The radiation scattered in the geometry of Figure 7.4 depends on the fluctuations in the electron density with wavenumber $\boldsymbol{k} = \boldsymbol{k}_s - \boldsymbol{k}_i$. The information about the plasma which this radiation yields depends on the parameter $\alpha = k\lambda_D$. If $\alpha \ll 1$ then the scattering from fluctuations on a scale much less than the Debye length is being measured. In this case the space charge shielding is unimportant and the scattering is just that from individual electrons. The fluctuations are due to the presence of a random distribution of discrete particles, and the power radiated is just the sum of that radiated from individual electrons in the classical Thomson scattering process. If an electron has velocity $\boldsymbol{v}$, the radiation scattered from it is Doppler shifted in frequency by an amount $\boldsymbol{k}\cdot\boldsymbol{v}$ so that the spectral width of the scattered radiation can give information on the temperature of the plasma.

If $\alpha \gtrsim 1$, then the scattering is from fluctuations of the size of the Debye length or longer. These correspond to the cloud of shielding electrons around each ion, and the spectral width of the scattered radiation now depends on the velocity distribution of the ions. In addition there are fluctuations at frequencies around ω_p corresponding to plasma oscillations, producing peaks in the scattered radiation separated from the central frequency by about this amount. Higher than thermal levels of ion sound or other waves give enhanced scattering of radiation shifted by the relevant wave frequency. The regime in which $\alpha \gtrsim 1$ is called collective scattering, since it comes from collective fluctuations in electron density, rather than individual electrons. It is an important technique for the investigation of plasma instabilities.

With a well-collimated incident beam and a narrow angle detector, the scattering volume indicated in Figure 7.4 may be very well localized. In laboratory-scale plasmas this can be achieved by using a laser as the radiation source. Laser scattering is a very useful way of obtaining spatially well-resolved information on plasma temperatures. In larger-scale plasmas, such as those of the ionosphere, longer-wavelength sources may be used in the same way. An example of the use of radiation scattering to probe the ionosphere is provided by the EISCAT (European Incoherent Scatter) project. In this the transmitter is at Tromso in northern Norway, with receivers there and at Kiruna in Sweden and Sodankyla in Finland. This system is designed to obtain detailed measurements on the ionosphere in the high-latitude regions in which the aurora occurs.

In the general class of techniques involving radiation sent into the plasma we may include interferometry. If a wave propagates through a plasma, then its shift in phase, compared to a wave going the same distance in a vacuum, is a measure of the integral of $n - 1$ along its path with n the refractive index

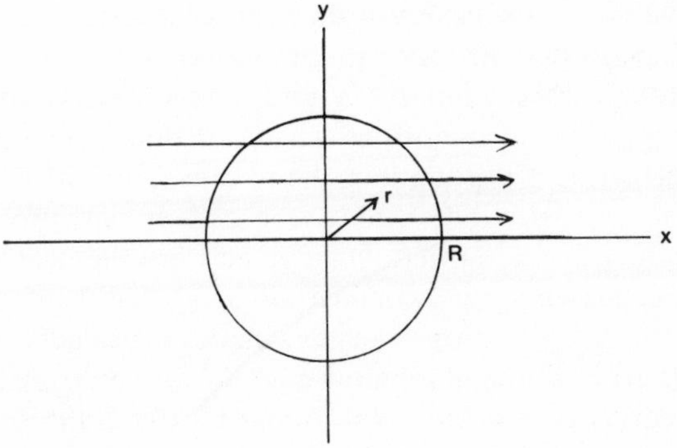

Figure 7.5 Interferometric measurement on a cylindrically symmetric plasma to yield refractive index by Abel inversion.

of the plasma. The shift in phase can be measured from the shift of interference fringes in an interferometer. If the plasma has suitable geometrical symmetry, space-resolved measurements of the refractive index may be obtained by a process known as Abel inversion. If we consider a cylindrically symmetric plasma, as shown in Figure 7.5, with refractive index $n(r)$, then a measurement on a ray passing through the plasma at height y gives

$$F(y) = \int_{-(R^2-y^2)^{1/2}}^{(R^2-y^2)^{1/2}} [n((x^2+y^2)^{1/2}) - 1]\,\mathrm{d}x = 2\int_y^R \frac{[n(r)-1]}{(r^2-y^2)^{1/2}}\, r\,\mathrm{d}r.$$

This is an integral equation for $n(r)$ of a type first studied by Abel, and its solution is

$$n(r) - 1 = -\frac{1}{\pi}\int_r^R \frac{\mathrm{d}F(y)}{\mathrm{d}y}\frac{\mathrm{d}y}{(y^2-r^2)^{1/2}}.$$

In a magnetized plasma the phenomenon of Faraday rotation may be used to acquire information about the magnetic field. If we consider waves propagating along the magnetic field, then from the discussion in Chapter 4 we see that the left and right circularly polarized waves propagate at different speeds. If plane polarized light is incident on the plasma the result is a rotation of the plane of polarization through an angle which depends on the magnetic field strength.

7.5 Particle measurements

Measurements on particles emitted from a plasma also provide information about it. Charged particle energies can be inferred from their deflection in known magnetic or electrostatic fields and various types of detector are based on this principle. They may be used in various situations, for instance to analyse the plasmas flowing outwards from a laser-irradiated target or to make measurements of ionospheric plasmas from satellites. In a magnetically confined plasma the charged particles are not directly accessible to detectors, but it is then possible to obtain information from neutral particles. Low-energy neutral particles drifting into the plasma from the wall region may undergo charge-exchange interactions with high-energy ions in the plasma. Such reactions involve a transfer of charge between the neutral and the ion, producing a high-energy neutral and low-energy ion. Measurement of the energy spectrum of these charge-exchange neutrals provides information on the energy spectrum of the plasma ions.

Just as was the case for radiation diagnostic techniques it is possible to use a more active approach and look at the attenuation of a beam of neutral particles injected into the plasma. Fast neutrals in the beam are absorbed through charge-exchange interactions with plasma ions and also through ionizing collisions with electrons or ions. They may also be scattered out of the beam without being ionized. From the total attenuation it is possible to obtain information on the plasma density.

Finally, in a plasma of sufficiently high temperature, the neutron yield from thermonuclear reactions gives an indication of the ion temperature. Since most reactions are due to the high-energy part of the ion energy distribution the reaction rate is very sensitive to the presence of a high-energy tail on the ion distribution or to beams of ions. Care must be taken to ensure that such effects do not lead to erroneous estimates of the bulk ion temperature.

Problems

7.1 Assuming that the radiation scattered by an electron at rest is of the same frequency as the incident radiation, show that the scattered radiation from a moving electron is shifted in frequency by $\boldsymbol{k}\cdot\boldsymbol{v}$, as stated in section 7.4.

7.2 If right and left circularly polarized waves of frequency ω propagate along a magnetic field with wavenumbers k_R and k_L respectively, show that the plane of polarization of linearly polarized radiation is rotated through $\frac{1}{2}(k_R - k_L)$ radians per unit length.

8 Application to fusion and space research

8.1 Introduction

The major areas of application of plasma physics are to research into controlled nuclear fusion and to space physics, and in this chapter we shall discuss some applications to these topics. Neither the list of topics discussed nor the treatment of individual topics is by any means exhaustive. The aim is simply to give some flavour of the range of subjects to which plasma physics is relevant.

Interest in ionized gases began in the late nineteenth century with experiments on gas discharges, though the term 'plasma' was not coined until the 1920s. It was introduced by Langmuir in the course of his study of electrostatic plasma waves. Until the beginning of the nuclear fusion programme, around 1950, laboratory work was confined to comparatively low-temperature gas discharges, and much of the impetus for the development of plasma physics came from space scientists.

In the period 1925–27 Appleton and his collaborators discovered the existence of the electrically conducting layers in what became known as the ionosphere. Much of the theory of wave propagation in cold plasmas, as given in Chapter 4, was developed soon after this to help explain the propagation of radio waves in the ionosphere. In the days before communications satellites, reflection of radio waves from the ionosphere was of vital importance for long-distance radio transmission.

At about the same time astrophysicists realized the importance of plasma physics for their subject and began the study of the kinetic theory of plasmas and of magnetohydrodynamics. The classic work of Alfvén on the waves which are now named after him was published in 1942.

The idea of nuclear fusion as a possible source of energy arose soon after the end of World War II and research was begun in a number of countries under conditions of secrecy. Gradually these conditions were relaxed until at the Second United Nations International Conference on the Peaceful Uses of Atomic Energy, held in Geneva in 1958, results of the various programmes were made public and open international cooperation on magnetic confinement began. This was no doubt prompted by the realization that initial optimism about the prospects for energy generation were unfounded and that any economic benefit from fusion would come only after a long research programme. From early experiments it became apparent that it was by no means a simple matter to confine a plasma with a magnetic field. Even if fluid instabilities could be suppressed there remained micro-instabilities which led to energy transport across the confining field at a rate much above that expected from classical transport theory. A great deal of effort has been devoted to the study of plasma stability as a result, and efforts to gain a better understanding of the basic properties of a plasma have also benefited space science.

An important turning point in nuclear fusion research occurred around 1968 when the results of the tokamak experiments carried out in the USSR were first announced. These showed confinement which was very significantly better than any previous experiment, and led to the building of many other tokamaks around the world. This type of machine is still the front-runner so far as magnetic confinement is concerned. At present research is at a particularly interesting stage since the large tokamaks TFTR (Tokamak Fusion Test Reactor) in the United States and JET (Joint European Torus), a joint venture by the countries of the European Economic Community plus Sweden and Switzerland, came into operation at the end of 1982 and the middle of 1983 respectively. These, and similar projected machines in Japan and the USSR, are designed to investigate the behaviour of plasmas in conditions reaching or approaching those needed for fusion. Their operation will be discussed in a little more detail in the next section.

Around 1970 previously classified calculations relating to fusion produced by laser compression of small target pellets were published. Since then interest in this concept of inertial confinement has stimulated a great deal of research on laser–plasma interactions. More recently considerable attention has been devoted to other methods of heating and compressing the target, using particle beams of various sorts. Further discussion of inertial confinement may be found in section 8.4.

Even if nuclear fusion proves to be a viable method of power generation, even on the most optimistic projection, it will not be until well into the twenty-first century that it makes any significant contribution. Its main advantage is the very abundant fuel resources. Deuterium is present as about one part in 10^4 of naturally occurring hydrogen and so there are vast reserves of it in the oceans of the world. As pointed out in Chapter 1 the reaction which can most readily be produced involves tritium, which is an unstable isotope of hydrogen not found in nature. To produce tritium a reactor, whether employing magnetic or inertial confinement, will be surrounded by a blanket of lithium, which absorbs the neutrons produced by the fusion reactions and produces tritium through the reaction

$$n + Li^6 \rightarrow He^4 + T^3.$$

The isotope Li^6 constitutes around 7% of natural lithium. Thus, tritium may be recycled within the reactor, and the fuel required in addition to deuterium is lithium, of which considerable reserves are also available. Heat generated in the lithium blanket by the high-energy neutrons will be used to drive generators in a conventional way.

Apart from the scientific and technological problems to be faced in the development of a fusion reactor, cost would seem to be a factor militating against its acceptance while other methods of power generation are available. Most projections seem to estimate the cost of a fusion power-generating plant to be two or more times that of a fission plant. The reason lies in the large volume of engineered reactor vessel which is required, because the absorbing blanket must be outside the volume where the reaction is taking place and must be big enough that the power flux through the wall can be handled. In a fission reactor coolant can surround a large number of individual fuel rods and give a much larger surface-to-volume ratio.

Although this book deals mainly with the properties of hot, completely ionized plasmas, relevant to fusion and to space physics research, we should also mention that there is important work being carried out on lower-temperature partially ionized plasmas in arcs and discharges. These studies are relevant to practical applications such as lighting systems, high-power switching devices or plasma arc furnaces. Plasma chemistry is also a field of growing importance, the reaction rates of many processes being considerably enhanced in a plasma environment.

8.2 Tokamaks and other toroidal devices

As was shown in Chapter 2 containment of a plasma in toroidal geometry requires that the magnetic field have both toroidal and poloidal components, with the latter being provided in a tokamak by a current flowing in the plasma, and the former by external coils. The rotational transform must be less than 2π, otherwise the Kruskal–Shafranov stability condition (see Chapter 3) is not satisfied. This condition is generally expressed in terms of the safety factor q, which is 2π divided by the rotational transform. Thus q must be greater than one everywhere within the plasma. For a given toroidal field this limits the poloidal field and hence the current in the plasma.

The plasma current in a tokamak is generally driven by having the plasma as the secondary coil of a transformer system, as shown schematically in Figure 8.1. An increasing current in the primary winding drives the plasma current. Such a system requires pulsed operation of the tokamak, and a considerable amount of research has been done recently on non-inductive current drive using neutral beams or radio-frequency waves to drive a current continuously. This possibility has been demonstrated on a number of devices, the PLT tokamak at Princeton having had a discharge maintained for several seconds solely with lower hybrid-frequency radiation. However the efficiency

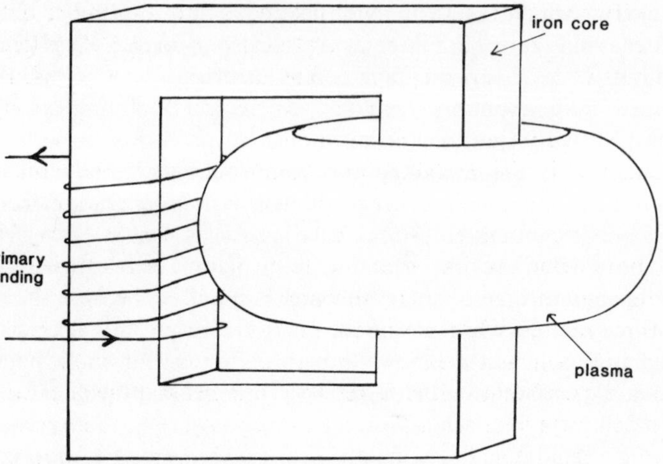

Figure 8.1 Circuit for driving plasma current in a tokamak.

of current drive processes, measured as the power required to drive a given current, is rather low and in a reactor a large fraction of the output would have to be recycled to maintain the current.

The current also heats the plasma, but because of the reduction in plasma resistivity with temperature, going as $T^{-3/2}$, and the fact that the current is limited by stability considerations, the temperature which can be reached by this ohmic heating is limited to 2–3 keV. To reach the higher temperatures needed for fusion some auxiliary heating system is required, either neutral beam injection or radio-frequency heating.

One of the main obstacles to reaching high temperatures in a tokamak is the presence of high-atomic-number impurities which greatly increase radiation losses. To minimize impurity concentrations the plasma must be kept away from the vessel wall. This is usually done by a limiter, which is simply an obstacle projecting into the plasma to intersect the outer field lines and produce a plasma boundary away from the wall. The material of the limiter may be chosen so as to minimize impurity production. An alternative is to

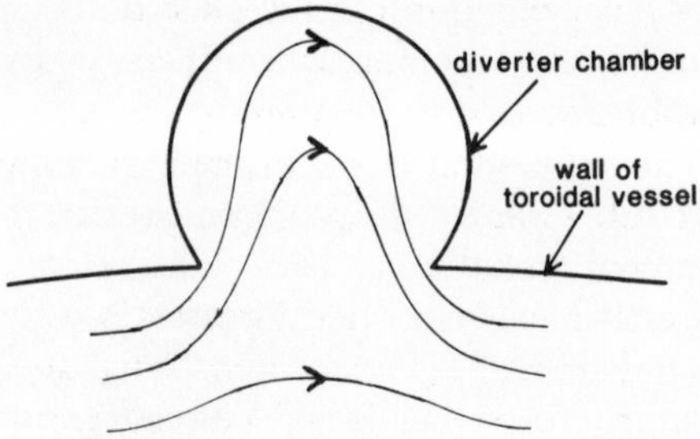

Figure 8.2 Magnetic field in a diverter.

have a diverter which takes the outer field lines into a chamber outside the main vessel as in Figure 8.2. The plasma associated with these field lines is then pumped out of the diverter chamber. The diverter is technologically more complicated than a limiter, but may be necessary if limiters introduce unacceptably high impurity concentrations.

As mentioned in the introduction a number of large tokamaks designed to approach reactor conditions have recently come into operation or are projected to come into operation within the next few years. We shall describe in some more detail the JET machine, to give some idea of the parameters of these devices. In this tokamak, the major radius of the plasma column is around three metres, while the minor cross-section of the vacuum vessel is D-shaped with a height a little over four metres and a width of 3 m. The D-shape is chosen because it is better able to withstand mechanical stresses, but should allow a somewhat higher plasma pressure than a circular cross-section. The toroidal magnetic field is around 3 T, while the plasma current is several MA.

Additional heating is planned to be added in stages during the years 1984–87. This will come from both neutral beam injection and radio-frequency heating in the ion cyclotron range of frequencies. Eventually some 25 MW of additional heating will be provided, which should be compared with the ohmic heating input of around 2MW and which should raise the plasma temperature to around 5 keV.

Initial operation will be with hydrogen or deuterium, but it is envisaged that in the late 1980s experiments will be carried out with a deuterium–tritium plasma. Once this is done the structure of the device will become radioactive, and any modifications or repairs must use remote-handling techniques.

The previous generation of tokamaks were typically of major radius litttle more than one metre, so a device like JET represents a considerable step upwards in size towards a reactor which might be about two or three times as large again. Plasma theory is not sufficiently well understood to enable one to say with confidence how results on a small machine will scale up to a larger device, and the main aim of these large machines is to investigate plasma confinement in conditions close to those required for fusion.

In Chapter 1 we have sketched the argument leading to the Lawson condition, necessary for a net gain of power from a reactor. A somewhat higher value of $n\tau$ may lead to ignition, where the α-particles (helium nuclei) produced by the fusion reactions and confined by the magnetic field have sufficient energy to maintain the plasma temperature without additional heating. Only if JET fulfils the most optimistic predictions of its behaviour will it reach ignition conditions.

Other toroidal devices have been overshadowed somewhat by tokamaks, but research on them continues nevertheless. The stellarator, which has a history going back to the work of Spitzer at Princeton in the early 1950s, uses external helical windings instead of a plasma current to create the necessary poloidal field. Production of the field in this way produces a somewhat different geometry from that of the tokamak. The field is no longer axisymmetric around the toroidal direction, but has a periodicity the same as that of the helical windings. For a reactor this type of device would have the advantage of being readily capable of steady-state operation, since no plasma current is necessary for stability.

Some interest has attached recently to the reverse-field pinch, a system in which the plasma current is well above the Kruskal–Shafranov limit, with stability being produced by strong shear of the field. The poloidal and toroidal fields are comparable in magnitude and there is a stable state in which the toroidal field in the outer part of the plasma column reverses its direction compared with the field in the centre of the discharge.

This by no means exhausts the list of different toroidal experiments, but the investment in tokamak research is such that it seems unlikely that any other system will take over as the most likely candidate for a fusion reactor based on magnetic confinement.

8.3 Other magnetic confinement systems

The best-known magnetic confinement device other than the toroidal ones is the magnetic mirror, whose principles of operation were discussed in Chapter 2. It is possible to create a mirror field which is MHD-stable, but particle losses through the mirrors lead to non-Maxwellian distributions and consequent mirror instabilities. A simple mirror has unacceptable end losses, but research continues, mainly in the United States, on the tandem mirror. This has a central mirror cell, with a number of smaller mirror cells at each end. Neutral beam injection and electron cyclotron heating are used to create distributions in the end cells which produce potential barriers preventing

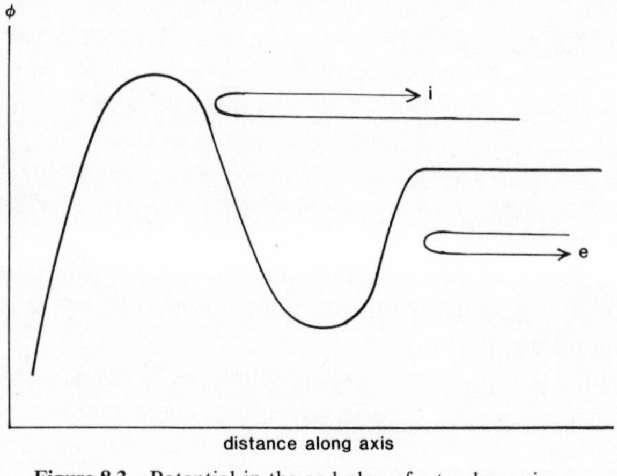

Figure 8.3 Potential in the end plug of a tandem mirror.

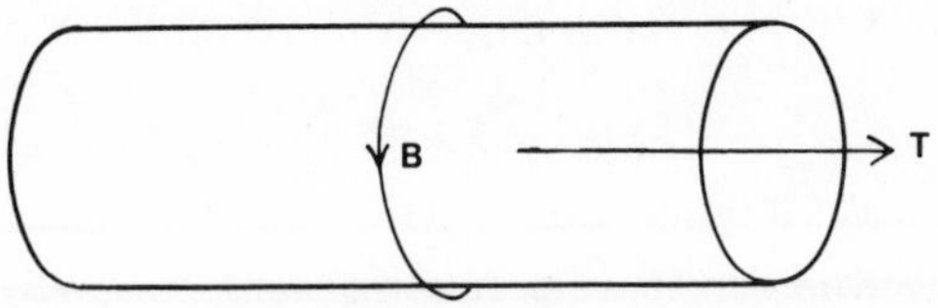

Figure 8.4 The Z-pinch.

particle loss, as shown in Figure 8.3. The effectiveness of such end plugs has been demonstrated, but whether the system has the potential to be developed into a reactor remains to be seen.

Another system which has been proposed as the basis for a reactor is the Z-pinch. This is simply a cylindrical current carrying plasma, confined by its own magnetic field, as illustrated in Figure 8.4. It would be produced by passing a discharge through a pre-formed conducting channel in a dense gas, the channel being produced by an electron beam or laser beam. The compression resulting from the magnetic field would then produce a plasma with a density of the order of 10^{27}–10^{28} particles per cubic metre, as compared with around $10^{20}\,\mathrm{m}^{-3}$ in a tokamak. With this high density, losses to the ends would take sufficiently long for fusion conditions to be feasible. It is necessary to avoid rapid break-up of the plasma due to kink or sausage instabilities, but it is thought that the dense neutral gas surrounding the plasma and the fact that the ion Larmor radius is comparable to the pinch diameter should both have a stabilizing influence. A fusion reactor utilizing this principle might have a pinch diameter of the order of 1 mm and a length of around 0.1 m. By its very nature it would be pulsed in operation, and the concept has much in common with the ideas of inertial confinement which we turn to next.

8.4 Inertial confinement

As explained briefly in Chapter 1 the basic idea of inertial confinement is that the confinement time is just the time that a small compressed target takes to fly apart. For a net fusion gain Lawson's criterion must be attained, but in this scheme the densities are many orders of magnitude higher than in a magnetically confined system and the containment time correspondingly shorter.

The need to compress the target to as high a density as possible may be understood from quite simple arguments. The containment time τ of a pellet of radius r is of the order of the time taken for a rarefaction wave moving at the speed of sound to travel a distance r, so

$$\tau \approx r/v_s. \tag{8.1}$$

The energy needed to heat the pellet to a temperature T is

$$E \approx 3n\kappa T\frac{4\pi r^3}{3}, \tag{8.2}$$

where n is the electron density. The required temperature is, as usual, about 10 keV, so that there is a reasonable reaction cross-section, and given this the sound speed is not strongly dependent on density, so that from (8.1) we see that Lawson's criterion can be expressed as a minimum value for nr. If this value is reached the energy required to heat the pellet to the required temperature can be seen from (8.2) to vary as r^2 or as n^{-2}.

At normal solid densities of a deuterium tritium target the pulsed energy required to heat it to fusion temperatures is impossibly large. Even if it could be done, the explosive output of fusion energy would be correspondingly large, creating problems with its absorption through the target chamber walls. Proposed schemes for inertial confinement fusion thus involve compression of the target, to several thousand times the solid density. Some energy is of course required to effect the compression, but the overall energy requirement is reduced to a level which is within possible reach of laser or particle-beam drivers.

In order to compress the target the most effective technique is to utilize ablative compression in which material is ablated from the target surface and the resulting reaction produces a compression wave moving inwards, as illustrated in Figure 8.5. The effect is often referred to as being like a spherical rocket, and the net effect is to blow off the outer layers of the target and leave a small compressed core.

Regardless of the nature of the incident beam driving the compression, there are some general requirements and problems associated with this process. In order that the compression be most effective the central part of

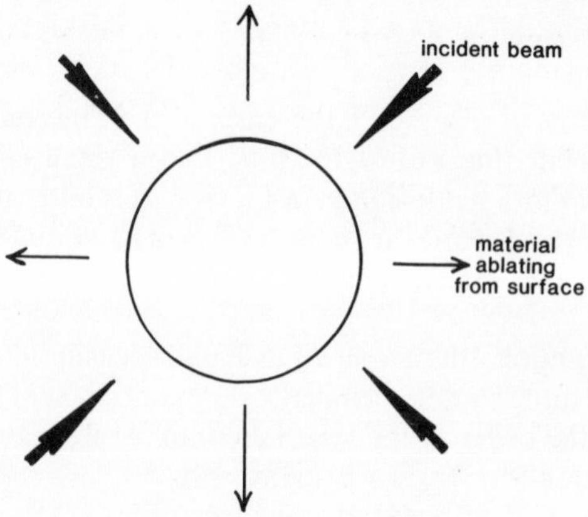

Figure 8.5 Ablative compression.

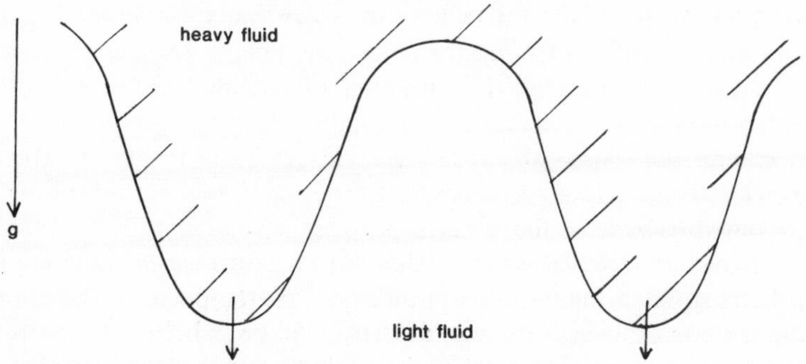

Figure 8.6 Rayleigh–Taylor instability.

the target must be as cool as possible. If, for example, the absorption process generates fast electrons which run ahead of the compression wave and preheat the core, the effect is to make the core more difficult to compress. Ideally the core should be compressed adiabatically to its final density since this requires the minimum energy expenditure.

Targets generally consist not of solid spheres, but of hollow shells, implosion of which gives a higher compression. In a fusion reactor the target would probably be layered, with a shell of heavier material surrounding the fuel. The inertia of the heavier material helps to compress and contain the core of the fuel. This type of system yields problems with the Rayleigh–Taylor instability. In its classical form this is the instability which occurs when a heavy fluid is supported against gravity by a lighter fluid. Any perturbation of a plane interface grows and leads to spikes of the heavier fluid penetrating into the lighter one as shown in Figure 8.6. In a compressed target the acceleration of the system towards the centre is equivalent to a gravitational force acting outwards and the Rayleigh–Taylor instability may lead to break-up of the shell. Suppression of this instability requires that the beam driving the compression should be as near uniform as possible and that the shell surface should be smooth, so as to minimize the initial amplitude of perturbations. Even with these precautions the existence of this instability places a lower limit on the ratio of shell thickness to radius and is a serious obstacle to effective compression of targets.

The driver for the compression may be either a laser beam or particle beams of various sorts. A laser-driven system consists of a number of beams focused on to a target of the order of 1 mm in diameter, arranged so as to make the radiation incident on the surface as uniform as possible. The duration of the pulse must be of the order of nanoseconds (1 ns $= 10^{-9}$ s). The choice of laser wavelength is of some importance. Laser light penetrates the target only up to the density at which the plasma frequency becomes equal to the light frequency. Energy must then be transferred by conduction from this critical density to the ablation surface. At longer wavelengths the distance between these surfaces is greater and there is more opportunity for smoothing of any irregularities in the incident beam before the energy reaches the ablation surface. At shorter wavelengths this thermal smoothing effect is less, but there is the advantage that penetration to higher densities produces a higher ablation pressure. Also, at shorter wavelengths absorption by collisions between electrons and ions is more effective. This inverse bremsstrahlung puts energy into the bulk of the plasma particles, avoiding production of fast particles which lead to preheating of the core. At longer wavelengths collisionless processes like resonant absorption (see Chapter 4) are more important, and characteristically produce a high-energy tail on the electron distribution function. Resonant absorption initially produces fast electrons directed away from the core, but the resulting excess of positive charge near the critical surface produces a potential which reflects them back inwards. This potential also accelerates ions outwards, adjusting itself to maintain approximate charge neutrality in the plasma. To achieve a balance between strong absorption by inverse bremsstrahlung and effective thermal smoothing the optimum wavelength seems to be around a third of a micron. Neodymium glass lasers used in laser compression experiments produce light at 1.06 μm, and recently the tendency has been to carry out experiments using the second or third harmonic of this.

One obstacle to the use of lasers for inertial confinement fusion is that the efficiency is rather low, which means that the energy gain of the target must be correspondingly high. However, research into the interaction of high-power lasers with targets has stimulated a great deal of work on non-linear plasma physics processes, some of which are discussed in Chapter 6. Also, while compression still falls far short of that required for a fusion reactor, present-day experiments can produce substantial compression. This allows the investigation of high-density plasmas which are non-ideal, that is, do not have a large number of particles in the Debye sphere, and opens up interesting areas of atomic physics and statistical mechanics. Even if lasers do not prove to be suitable for driving inertial fusion, the study of laser–plasma interactions has produced a great deal of interesting plasma physics and improved our understanding of nonlinear processes.

For inertial confinement driven by particle beams, electrons and both heavy and light ions have received consideration. The main requirement is that the particles must be stopped and deposit their energy within the target shell without reaching the fuel and preheating it. Scattering of electrons makes this difficult to achieve, and ions, which have a much more sharply defined range in the target, are a better prospect. The energy of a proton absorbed in a suitable target thickness would typically be around 10 MeV, while that of a uranium ion would be around 10 GeV. To have the same power deposition would require a much lower flux of the heavy ions. Typically the beam current in a light ion beam would be a few tens of MA, while that in a heavy ion beam would be a few thousand times less. The smaller current means that space charge effects, and focusing difficulties because of the mutual repulsion of particles, are much less severe. Focusing of the beam on to the target, a matter of using conventional optics for a laser, is in fact one of the main problems to be faced in particle beam fusion. Present studies indicate that heavy ion fusion is perhaps the most promising method of inertial confinement. It presents fewer problems with focusing than light ion fusion, and the accelerator technology developed for nuclear and high energy particle physics should allow generation of pulses with high efficiency and the required high repetition rate. The higher efficiency of ion accelerators, perhaps around 25% compared with at most around 5% for laser systems, means that a lower gain is required in the target and a lower energy in each pulse.

8.5 The Earth's magnetosphere

The dipole magnetic field of the Earth interacts with the solar wind, a stream of plasma originating in the Sun, to produce a configuration rather like that shown in Figure 8.7. The solar wind has a supersonic flow velocity and produces a shock wave when it meets the magnetosphere. This bow shock is similar in overall structure to the shock produced by any blunt body moving supersonically in a gas, but its detailed structure is extremely interesting and we shall say a little more about it shortly. On the downstream side of the Earth the magnetic field is drawn out into a long tail as shown, with a neutral sheet where the magnetic field changes its sign.

In the upper part of the atmosphere, the ionosphere, plasma is produced through the ionization of neutral atoms by the ultraviolet component of the Sun's radiation, while at higher levels most of the plasma originates in the solar wind. The radiation belts contain energetic particles, mainly protons

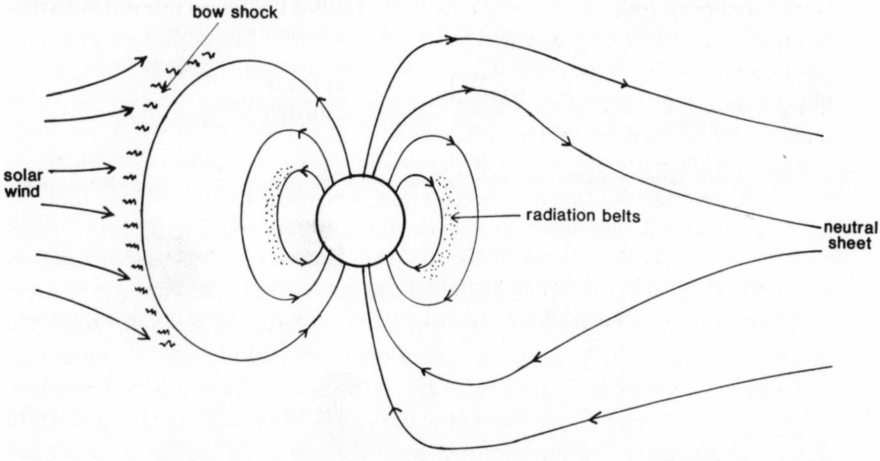

Figure 8.7 The magnetosphere.

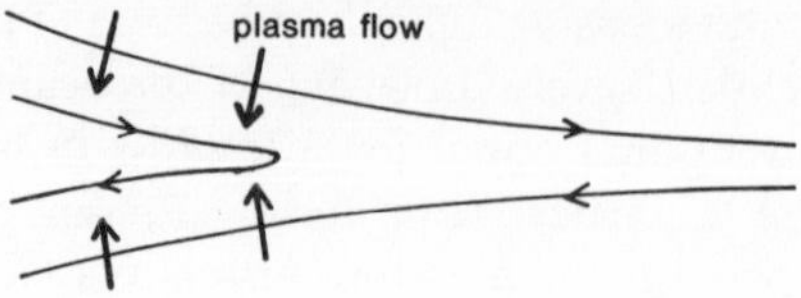

Figure 8.8 Reconnection in the neutral sheet.

and electrons, trapped in the dipole field. In this region collisions are sufficiently infrequent that particles can be trapped for considerable periods.

The neutral sheet is a region of some interest. In the plasma, conductivity is so high that over most of it the magnetic field is frozen in as discussed in section 3.2. However, this condition no longer holds when the magnetic field is small, as it is in the vicinity of the neutral sheet, on which it vanishes altogether. In such a region magnetic reconnection can take place with a field line following the plasma flow changing its topology from an open to a closed configuration as shown in Figure 8.8. This is accompanied by resistive dissipation of energy, driven by the stored energy of the magnetic field. This is thought to be the source of some of the magnetic disturbances which occur in the ionosphere.

The bow shock, which is about ten earth radii from the Earth, is an example of a plasma physics phenomenon known as a collisionless shock. The name derives from the fact that binary collisions between particles, of the sort discussed in Chapter 3, cannot provide the dissipation across the shock. Essentially the shock consists of a layer, less than 100 km thick, in which the solar wind is slowed to subsonic velocities and its kinetic energy converted to thermal energy. The particle mean free path in the solar wind is comparable with the distance from the Earth to the Sun, so ordinary collisional dissipation cannot explain the observed structure. Similar phenomena have been observed in the laboratory and have attracted a lot of research.

The bow shock is particularly interesting and complicated because in different parts of the shock front and at different times it displays a wide range of different conditions. Much depends on whether the magnetic field is perpendicular to the flow velocity of the incoming plasma, oblique or parallel, quite different structures being found in the different cases. Perpendicular or quasi-perpendicular shocks have the most regular structure which is quite easy to explain, at least qualitatively. They are also of the type which have been most thoroughly investigated in the laboratory.

The structure of such a shock is illustrated in Figure 8.9. In the rest frame of the shock the electric field in the y-direction is such that

$$v = \frac{\boldsymbol{E} \times \boldsymbol{B}}{B^2}, \tag{8.3}$$

that is the plasma is drifting with the $\boldsymbol{E} \times \boldsymbol{B}$ drift velocity discussed in

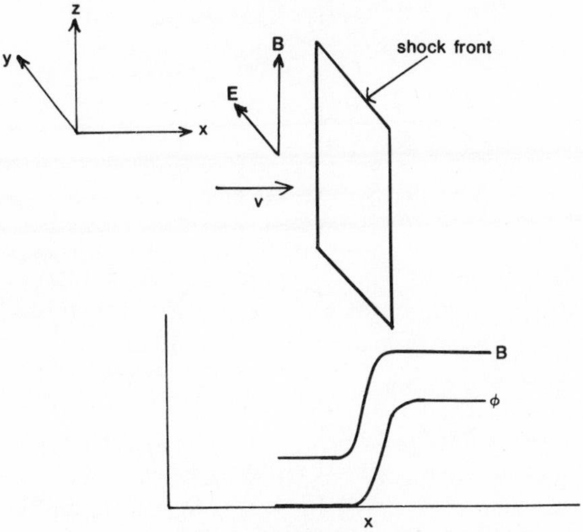

Figure 8.9 Perpendicular shock.

Chapter 2. The condition

$$\nabla \times \boldsymbol{E} = -\frac{\partial \boldsymbol{B}}{\partial t}$$

implies that in a steady configuration E_y is constant. Measurements on perpendicular shocks show that there are jumps in the magnetic field and electric potential through the shock as illustrated in Figure 8.9, though small-scale time-dependent fluctuations are superimposed on this overall structure.

The structure of the shock may be understood if it is noted that its thickness is large compared to an electron Larmor radius, so that electrons follow guiding centre orbits of the type discussed in Chapter 2, but small compared to the ion Larmor radius. According to (8.3) the effect of the increase in magnetic field in the shock front is to slow down the electron flow in the x-direction. However, over a scale length shorter than the Larmor radius the same mechanism cannot slow the ions. To maintain quasi-neutrality a potential jump develops in order to slow the ions by the same amount as the electrons and so keep their densities equal. The electric field E_x associated with this potential jump produces a drift of the electrons in the y-direction, and the associated current is just in the right direction to produce the increase in magnetic field. Dissipation within the shock is provided by instabilities excited by the electron drift with respect to the ions in the y-direction. These produce an enhanced level of fluctuations and effective resistivity well above that due to ordinary collisions. In this way we can see, in outline, how a self-consistent structure of the type described above can be set up, though making detailed quantitative predictions is by no means so easy. If the ions have a sufficient

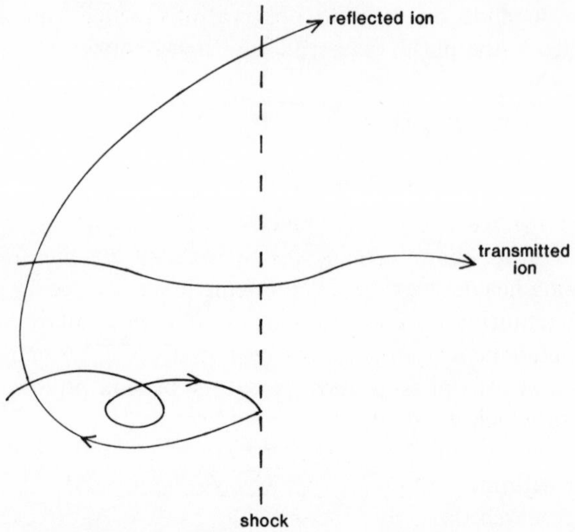

Figure 8.10 Ion orbits through a perpendicular shock. Most ions are transmitted through the shock, but slow ions may be reflected first.

thermal spread in their upstream velocity distribution then some of the more slowly-moving ions may be reflected by the potential in the shock front. These are bent around and accelerated in the upstream fields and re-enter the shock with high velocity as indicated in Figure 8.10, producing a beam of fast ions in the downstream region, and a 'foot' in the magnetic field profile upstream. This provides a mechanism by which some of the incoming energy may be converted to ion thermal energy in the downstream region.

Oblique and parallel shocks are generally much thicker than perpendicular shocks and show much more random fluctuating structures. Theoretical understanding of such shocks is more limited, but dissipation seems to take place by scattering in a rather wide turbulent region. Before leaving the subject of shocks we should mention that strong collisionless shocks are thought to occur in the tenuous plasma in interplanetary and interstellar space, and it has been suggested that they are involved in the production of cosmic rays.

The bow shock provides a very interesting application of plasma physics to the understanding of the magnetosphere, and its study has been stimulated by very detailed satellite observations of its structure. It provides an ideal laboratory for the study of collisionless shocks, since its large scale makes the fine details of its structure much more accessible than in laboratory shocks.

There are also many other plasma physics problems involved in understanding the magnetosphere and its interaction with the solar wind, in order to explain such phenomena as the aurorae and magnetic storms. A great deal of information about the lower magnetosphere can be obtained by ground-based radio observations and by balloon or rocket sounding. Study of the more remote parts depends on satellite observations which can give detailed information about the plasma surrounding the satellite.

8.6 The physics of the Sun and stars

The Sun and stars are successful nuclear fusion reactors, deriving their energy from a rather complicated series of reactions starting from hydrogen and synthesizing heavier elements. Confinement is provided by gravitational forces which, requiring an enormous mass, are not available to terrestrial attempts at confinement. There are, however, many points of contact between solar physics and astrophysics and laboratory plasma physics and nuclear fusion, some of which we shall touch upon here.

The central part of the Sun and stars where the reactions take place is a dense plasma with more in common with the compressed targets of inertial confinement research than with magnetically confined plasmas. Problems of radiation transport and the behaviour of dense ionized matter are common to astrophysics and to inertial confinement, but we shall not discuss them further since they are more concerned with atomic physics and spectroscopy than with plasma physics.

Plasma physics of the type discussed in this book is more relevant to the upper atmosphere of the Sun. Three layers are generally distinguished. Lowest is the photosphere which is dense and opaque and is the source of most of the visible radiation. Its temperature is around 6000 K. Above it is the chromosphere which is slightly cooler, rather less dense and transparent, while above that again is the corona in which the temperature rises to about 10^6 K and the density drops once more. Electron densities are of the order of $10^{21}\,\mathrm{m}^{-3}$ in the photosphere, dropping to $10^{15}\,\mathrm{m}^{-3}$ in the chromosphere. The corona has no well-defined outer edge, extending to the Earth's orbit and beyond in the form of the solar wind.

The outward flow of material in the corona which constitutes the solar wind was first directly observed by spacecraft in 1959, but its existence had been postulated a few years earlier. The main reason was that an acceptable structure does not arise from steady-state equations, solutions of which have a pressure which does not tend to zero at large distances from the Sun. The following solution incorporating a steady outward flow of matter was given by E.N. Parker. Assuming a spherically symmetrical steady flow, conservation of mass gives

$$4\pi r^2 \rho v = \text{constant}, \tag{8.4}$$

and the momentum equation is

$$\rho v \frac{\mathrm{d}v}{\mathrm{d}r} = -\frac{\mathrm{d}}{\mathrm{d}r}(p) - \frac{GM\rho}{r^2} \tag{8.5}$$

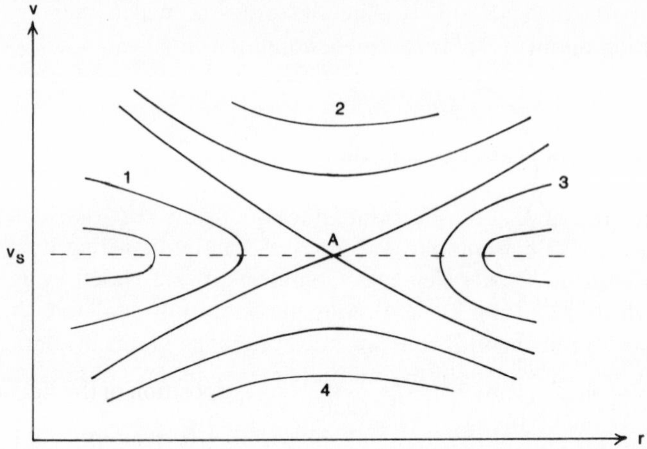

Figure 8.11 Solution curves of the differential equation 8.6.

The last term here is the gravitational attraction of the Sun, with M the solar mass and G the gravitational constant. Assuming the temperature to be constant we have a constant sound velocity given by $v_s^2 = p/\rho$. Eliminating ρ from (8.4) and (8.5) gives

$$\left(v - \frac{v_s^2}{v}\right)\frac{dv}{dr} = \frac{2v_s^2}{r} - \frac{GM}{r^2} \tag{8.6}$$

which can be integrated to give

$$\left(\frac{v}{v_s}\right)^2 - \log\left(\frac{v}{v_s}\right)^2 = 4\log r - \frac{2GM}{rv_s^2} + \text{constant} \tag{8.7}$$

The solution curves defined by (8.7) are as shown in Figure 8.11.

Regions 1 and 3 give solutions where v is a double-valued function of r and which are not physically acceptable. Region 2 has solutions which imply supersonic velocities at the Sun's surface, contrary to observation. Acceptable solutions are the completely subsonic solutions in region 4, or the solution which is subsonic near the Sun and becomes supersonic on passing through the critical point A. This latter solution is the one which is in agreement with observation.

In practice the solar wind is not as simple as this and more complicated calculations can be done taking into account magnetic fields, temperature variations, departures from spherical symmetry and so on. However, the basic fact that material is streaming outwards from the Sun is well established. The form of the magnetic field convected outwards with the solar wind can be seen qualitatively from quite simple considerations. Assuming that a field line is tied to a fixed point at the surface of the Sun and that it is frozen into the plasma flow, as discussed in section 3.2, then the field line will trace out the

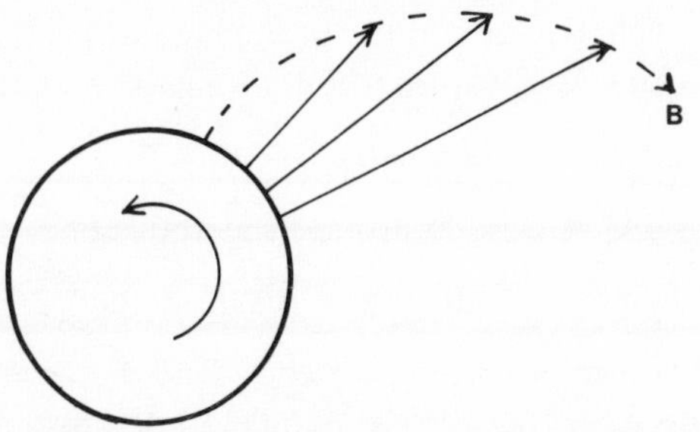

Figure 8.12 Spiral field lines produced by rotation of the Sun.

path of material ejected from that fixed point. The rotation of the Sun makes this into a spiral as shown in Figure 8.12. At the position of the Earth the angle between the field and the flow is about $\pi/4$.

The mechanism by which the magnetic field of the Sun is generated is itself an interesting problem. The conductivity of the Sun is such that the time over which the field decays is much shorter that the Sun's lifetime, indicating that there must be some dynamo mechanism to maintain the field. The dynamo problem is usually divided into two parts, the first, the kinematic dynamo problem, being to find fluid motions which, substituted in equation (3.9), i.e.

$$\frac{\partial \boldsymbol{B}}{\partial t} = \frac{\eta}{\mu_0} \nabla^2 \boldsymbol{B} + \nabla \times (\boldsymbol{v} \times \boldsymbol{B})$$

will maintain the field against the dissipative effect of the first term on the right-hand side. Having identified possible classes of motion, the second part of the problem is to explain how these might be produced.

Various anti-dynamo theorems show that dynamo action producing steady fields with a high degree of symmetry is not possible. The best known is that of Cowling which shows that a steady axisymmetric field cannot be maintained.

The argument can be understood by referring to Figure 8.13. Such a field must have a line, like the dotted line in the diagram, on which the poloidal

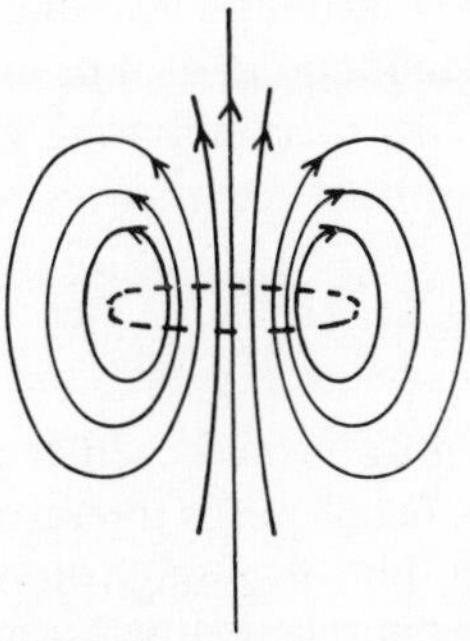

Figure 8.13 A steady axisymmetric magnetic field.

component of the field (i.e. that in the (r, θ) plane with spherical coordinates chosen in the obvious way) vanishes. Ohm's law says that

$$\boldsymbol{J} = \sigma(\boldsymbol{E} + \boldsymbol{v} \times \boldsymbol{B}),$$

and if this equation is integrated around the above line,

$$\oint \boldsymbol{E} \cdot \mathrm{d}\boldsymbol{l} = \oint (\boldsymbol{v} \times \boldsymbol{B}) \cdot \mathrm{d}\boldsymbol{l} = 0.$$

The first integral is zero because in a steady field $\boldsymbol{\nabla} \times \boldsymbol{E} = 0$ and the second because on the line $\boldsymbol{B}$ is either zero or parallel to $\mathrm{d}\boldsymbol{l}$. This implies vanishing of the current around this line, which is incompatible with Ampère's law.

Much recent work has concentrated on turbulent dynamos, in which small-scale fluctuations in the velocity and field maintain a non-zero average field. The basic idea is to split $\boldsymbol{B}$ and $\boldsymbol{v}$ into slowly-varying averages $\boldsymbol{B}_0$ and $\boldsymbol{v}_0$ and fluctuations $\boldsymbol{B}_1$ and $\boldsymbol{v}_1$. Then the average of (3.9) gives

$$\frac{\partial \boldsymbol{B}_0}{\partial t} = \frac{\eta}{\mu_0} \nabla^2 \boldsymbol{B}_0 + \boldsymbol{\nabla} \times \langle \boldsymbol{v}_1 \times \boldsymbol{B}_1 \rangle, \tag{8.8}$$

the angular brackets denoting an average over the small-scale fluctuations. Subtracting (8.8) from (3.9) gives an equation for the fluctuations, the procedure being formally very similar to the quasilinear theory discussed in section 6.2. Analysis of this system of equations shows that a mean magnetic field $\boldsymbol{B}_0$ will be generated only if the motion has non-zero helicity, i.e. $\boldsymbol{v} \cdot (\boldsymbol{\nabla} \times \boldsymbol{v}) \neq 0$. This demands that the flow has a rotational component, together with a velocity along the axis of rotation, rather like water flowing out of a bath. One possible source of such a motion is the Coriolus force due to the Sun's rotation which might be expected to produce cyclonic motions just as it does in the Earth's atmosphere.

When looked at in detail the surface layers of the Sun are by no means uniform. The chromosphere contains convection cells on different scales forming the structure known as granulation and supergranulation. The magnetic field is not uniform, but is concentrated into localized flux tubes, perhaps as a result of convection. Sunspots, which are dark cool areas containing stronger than average magnetic fields, appear as transient features on the surface. They are often found in groups and are surrounded by regions with a complex magnetic field structure, known as active regions. Such groups may persist for several months. Prominences, which are projections of the relatively cool dense chromosphere into the corona, are associated with sunspots in the active regions and may also have a lifetime measured in months.

The apparent stability of prominences, which are presumed to be supported against gravity by magnetic fields, poses interesting problems in magnetostatics, the challenge being to find a configuration which looks like what is observed and is stable. Solar flares, which are events involving the rapid release of large amounts of energy in the form of radiation and fast particles, also occur in the active regions. The most likely source of the energy is magnetic energy, with a complex magnetic field structure slowly evolving until it becomes unstable, then producing a rapid release of energy. Various models have been suggested, the mechanism by which the field energy is released generally being reconnection in the vicinity of a neutral point in the field, as discussed in the last section.

As well as the magnetohydrodynamics of the large-scale structures and motions in the Sun, there are many problems which involve the more detailed microscopic behaviour of the plasma. Among these are the processes involved in particle acceleration in solar flares, and the mechanisms through which the fast particles generated give rise to radio bursts. The physics of the Sun and other similar stars thus has much in common with the physics of nuclear fusion devices and there is much scope for valuable cross-fertilization between the two fields.

8.7 Pulsars

Pulsars provide an example of astronomical objects in which very extreme conditions exist and which have a magnetosphere quite different from that of the Sun or Earth. These objects, which produce pulsed radiation signals with periods typically in the range 0.03–3s are generally agreed to be rotating neutron stars. These are stars which, as a result of gravitational collapse, have reached pressures such that the electrons and protons which form a large part of ordinary matter are combined to form neutrons, and the major constituent of the star is a Fermi-degenerate gas of neutrons, although enough electrons and protons remain to make it highly conducting. Such bodies, of about the same mass as the Sun but of the order of 10 km in radius, can rotate with the necessary high frequency, and the usually accepted explanation of the pulsed radiation is that it is produced by a lighthouse effect as shown in Figure 8.14.

Plasma physics is of relevance to the magnetosphere surrounding the

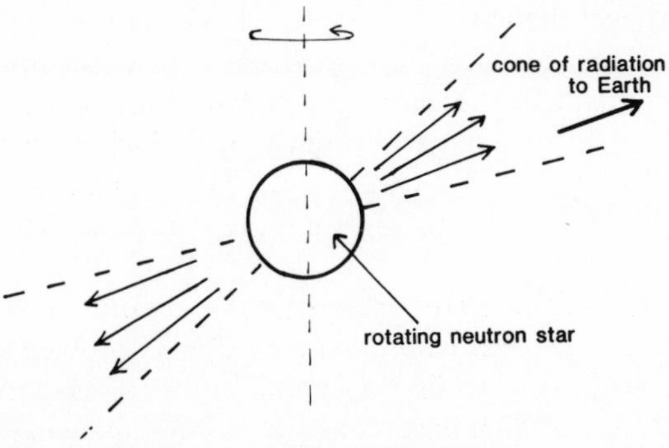

Figure 8.14 Lighthouse effect producing pulses of radiation.

neutron star and to the generation of the radiation. A complete theory of the processes occurring is not available, but some features which seem to be generally accepted will be described here. The main point to note is that magnetic fields frozen into the collapsing star produce very high field strengths at the surface of the star, estimated to be around 10^8T. Since the star is rotating and is a very good conductor there must be an electric field such that

$$\boldsymbol{E} + \boldsymbol{v} \times \boldsymbol{B} = 0$$

$$\text{i.e.} \qquad \boldsymbol{E} + (\boldsymbol{\Omega} \times \boldsymbol{r}) \times \boldsymbol{B} = 0, \tag{8.9}$$

with $\boldsymbol{\Omega}$ the angular velocity and $\boldsymbol{r}$ the position vector with respect to the centre of the star. At the surface of the star the electric field is 10^{10}–10^{12} Vm^{-1}. If there were a vacuum outside the surface this field would have a component along the magnetic field and its value at the surface would be enough to pull ions or electrons from the surface. It is also strong enough for there to be production of electron–positron pairs in the vacuum. The result is that the region outside the surface is populated with charged particles. Equation 8.9 then extends to the magnetosphere where the charge density is such as to neutralize the electric field along magnetic lines of force so that $\boldsymbol{E}\cdot\boldsymbol{B} = 0$. The charge density is given by $(1/\varepsilon_0)\,\boldsymbol{\nabla}\cdot\boldsymbol{E}$ and from (8.9) can be calculated to be

$$\rho = \frac{\boldsymbol{\Omega}\cdot\boldsymbol{B}}{2\pi\varepsilon_0}, \tag{8.10}$$

if the magnetic field due to the current flow in the magnetosphere can be neglected.

The picture so far is thus of a co-rotating magnetosphere with particles

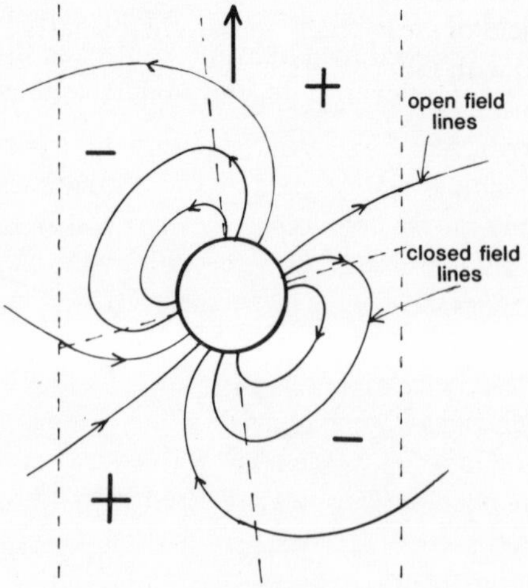

Figure 8.15 The magnetosphere of a pulsar.

carried around with the rotation of the star. It is generally supposed that the charges giving rise to the current density (8.10) are all positive or all negative as appropriate. However, there is a cylinder around the star on which $|\boldsymbol{\Omega} \times \boldsymbol{r}| = c$ and outside this the particles can no longer co-rotate with the star since the resulting velocity would exceed that of light. If it is assumed that particles flow along field lines, then outside the light cylinder the fields must have an azimuthal component. This requires some current flow along the field lines, so we have the picture shown in Figure 8.15. The closed field lines, that is those completely within the light cylinder, have stationary charges co-rotating as described, positive or negative in the regions indicated. Along open field lines there is a flow of charge, though just how or whether the overall charge in the star remains constant is not clear. Many problems connected with the structure of the pulsar magnetosphere remain to be solved.

Another major question is what produces the intense radiation. The most popular scheme involves charges flowing along the curved open field lines in the polar regions. An instability can produce bunching of the particles which in turn produces coherent radiation emission. This would be emitted in a cone around the polar regions and so produce the required lighthouse effect if the magnetic and rotational axes were inclined. Again, however, there is no universally accepted theory and many open questions remain.

8.8 Cosmic ray acceleration

As a final indication of the importance of plasma physics in astrophysics we give a brief discussion of a process which has been proposed for cosmic ray acceleration. It is similar in principle to the method originally proposed by Fermi which involved reflection from moving magnetic mirrors associated with the magnetic field of the Galaxy. Assuming the particle momentum to be negligible compared with that of the structure associated with the mirror, then the particle momentum perpendicular to the mirror is simply reversed in the rest frame of the mirror. In the frame of the observer the particle gains energy if the mirror is approaching it and loses energy if it catches up on a receding mirror. In a randomly moving assembly of mirrors, collisions of the first type are more probable than those of the second and the particle gains energy overall. This is an elegant idea, but it is difficult to identify moving structures in the Galaxy which would produce mirrors moving sufficiently rapidly to give the required particle acceleration.

Recently, however, it has been suggested that strong collisionless shocks moving through the Galaxy may be responsible for cosmic ray acceleration. The basic idea can be described with reference to Figure 8.16. Plasma is compressed as it crosses the shock, so in the rest frame of the shock conservation of matter tells us that $u_2 < u_1$. In fact in a high Mach number strong shock $u_2 \approx \frac{1}{4}u_1$. Now suppose we are in a frame of reference where the

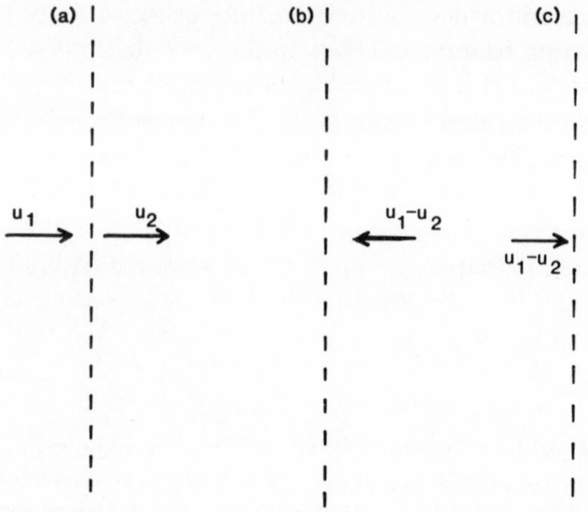

Figure 8.16 Shock in rest frames of (*a*) shock front, (*b*) upstream plasma, (*c*) downstream plasma.

upstream plasma is at rest and there is a distribution of high-energy particles. If one of these particles crosses the shock front into the downstream region, then, as is clear from Figure 8.16 (*b*), its momentum perpendicular to the shock front is in the opposite direction to the motion of the downstream plasma. In the rest frame of the downstream plasma it therefore has a higher energy than in the rest frame of the upstream plasma. Now suppose that there are fluctuations in the downstream plasma to scatter such particles and produce an isotropic distribution. Then a particle may be scattered back upstream. But again as it crosses the shock front its energy in the rest frame of the upstream plasma becomes greater than that it had in the rest frame of the downstream plasma and hence greater than it had originally in the upstream plasma. If there are fluctuations upstream to scatter the particles there, or another shock converging on the first, the process may continue.

Detailed analysis of this process involves consideration of the fluctuations in the neighbourhood of collisionless shocks and the way in which they scatter particles. For the mechanism to work particles have to be scattered in direction without losing too much energy.

8.9 Plasma physics applied to particle accelerators

This final example of the application of plasma physics, while it does not quite fall into the categories of the chapter title, is one which is currently attracting interest and so is worth describing briefly here. High-energy physicists are always seeking results at higher energies requiring more and more powerful accelerators. With conventional technology the accelerating electric field is limited to around 10^7 Vm^{-1} and so higher energies require longer acceleration lengths and ever bigger machines. As a result attention is being given to alternative accelerator designs, and one interesting concept to have arisen is that of the plasma beat wave accelerator.

The basic idea of this is to use the beating between two high-power laser beams to produce an electrostatic wave. The electromagnetic waves produced by the lasers have the dispersion relation

$$\omega^2 = \omega_p^2 + k^2c^2,$$

and if a difference in frequency $\Delta\omega$ corresponds to a difference in wavenumber Δk, then

$$\frac{\Delta\omega}{\Delta k} \approx \frac{\partial\omega}{\partial k} = c\left(1 - \frac{\omega_p^2}{\omega^2}\right)^{1/2}. \tag{8.11}$$

If $\Delta\omega$ is arranged to coincide with the plasma frequency in an ambient cold plasma (with $\omega_p \ll \omega$) then a plasma wave can be set up by the beats between the two laser beams, with a phase velocity given by (8.11). This phase velocity is just below c, so a particle can move along with the wave and be accelerated by its electric field. The limit on the acceleration is given by the fact that as it accelerates the particle moves out of phase with the wave, although it should be noted that for highly relativistic particles the rate of change of velocity with energy is small.

An alternative idea which avoids this phase slippage is to have a wave propagating perpendicular to a magnetic field, as shown in Figure 8.17. If the particle is trapped by the wave and pulled along in the x-direction with the wave speed then the $\boldsymbol{v} \times \boldsymbol{B}$ force acts in the y-direction and produces a steady acceleration. The particle need not remain exactly in phase with the electrostatic wave, but need only remain trapped, a condition which it is possible to satisfy.

In principle either of these methods yields acceleration distances much

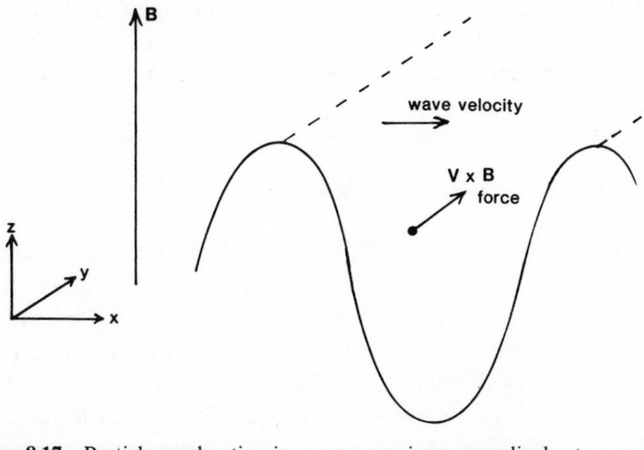

Figure 8.17 Particle acceleration in a wave moving perpendicular to a magnetic field.

shorter than those of conventional accelerators. However, these ideas are presently at a preliminary stage only, and much scientific and technological development will be necessary before an accelerator based on them becomes a reality, if indeed they prove to be viable at all.

8.10 Conclusion

This chapter gives a brief and somewhat impressionistic survey of some of the areas of laboratory and space physics in which the behaviour of plasmas plays an important role. It is by no means complete and many of the individual topics could themselves be made the subject of a book, but should give some idea of the range of application of plasma physics.

It should have become apparent that a plasma is a very complicated and by no means perfectly understood system. The history of confinement research is largely a history of finding out the hard way that plasmas are not as simple as was thought. Even in the early 1970s enthusiasts for laser-driven fusion were predicting that demonstration reactors would be operational in a few years. A cynic might be forgiven for suggesting that the number of years from the present to when it is projected that a demonstration reactor will be built, is an increasing function of time.

On a more optimistic note there is no doubt that great advances have been made in understanding plasma physics and that steady progress towards the goal of controlled nuclear fusion is being made. An improved understanding of the detailed behaviour of plasmas is also of vital importance in space and astrophysics, where the much more detailed observations made possible by the space programme require a more detailed theoretical treatment, involving not only the large-scale fluid motions of the plasma but the microscopic theories of instabilities and other processes which were found essential in understanding the behaviour of laboratory plasmas.

Appendix: Singular integrals

Most of the mathematics used in this book involves subjects, like vector analysis, which should be familiar to anyone with the requisite background in electromagnetism. A possible exception is the theory of singular integrals which is necessary for the section on Landau damping in Chapter 5, so here we shall give a brief account of the relevant properties.

We are concerned with integrals of the form

$$F(z) = \int_C \frac{f(t)}{t-z}\,\mathrm{d}t \qquad \text{(A.1)}$$

where C is some smooth contour in the complex plane which we take to be infinite since we do not wish to concern ourselves with effects at endpoints. The results which we state are dependent on $f(t)$ being a sufficiently well-behaved function on the contour. It suffices that it satisfy a Hölder condition, which states that for any two points on the contour t_1 and t_2

$$|f(t_1) - f(t_2)| < A|t_1 - t_2|^{\mu}$$

for positive constants A and μ.

Then equation A.1 defines $F(z)$ as an analytic function of z at any point which does not lie on the contour. If z lies on the contour then the principal part of the integral exists and is denoted by

$$P\int_C \frac{f(t)}{t-z}\,\mathrm{d}t.$$

This is defined by removing an interval of length 2ε, symmetrical around z, from the range of integration, then letting $\varepsilon \to 0$. Note that for the integral to be convergent in the usual sense at the singularity the endpoints would have to be capable of being allowed to tend to z from each side independently.

The properties of these integrals which are most frequently used in plasma physics are expressed in the Plemelj formulae. If $F^+(z)$ is the limit of $F(z)$ as z tends to the contour from the left-hand side (looking along the direction of integration) and $F^-(z)$ the corresponding value as x tends to the contour from the right-hand side, then

$$F^+(z) = P\int_C \frac{f(t)}{t-z}\,\mathrm{d}t + i\pi f(z)$$

$$F^-(z) = P\int_C \frac{f(t)}{t-z}\,\mathrm{d}t - i\pi f(z). \qquad \text{(A.2)}$$

Thus F is analytic in each of the two regions separated by the contour, but has a discontinuity on the contour. On going across the contour the value of $F(z)$ jumps by $2i\pi f(z)$.

The equations A.2 are sometimes expressed by saying that as z approahes the contour

$$\frac{1}{t-z} \rightarrow P\frac{1}{t-z} \pm i\pi\delta(z) \tag{A.3}$$

the $\pm$ depending on from which side z approaches the contour. This expression is only meaningful if $1/(t-z)$ is part of an integrand.

Further reading

General

A number of textbooks on plasma physics have been published, though many of the older ones are now out of print. Useful modern texts are:

F.F. Chen (1984) *Introduction to Plasma Physics and Controlled Fusion*, 2nd edn., Plenum, London.

D.R. Nicholson (1983) *Introduction to Plasma Theory*, John Wiley & Sons, Chichester.

K. Myamoto (1980) *Plasma Physics for Nuclear Fusion*, MIT Press, Cambridge, Mass.

R.D. Gill (ed.) (1981) *Plasma Physics and Nuclear Fusion Research*, Academic Press, London.

Below we give some more specialized references relevant to specific chapters of this book.

Chapter 1

The conditions necessary for fusion were given by J.D. Lawson (1957) *Proc. Phys. Soc.* **70B**, 6.

Chapter 2

A good discussion of single particle motion and many further references are given by P.C. Clemmow and J.P. Dougherty (1969) *Electrodynamics of Particles and Plasmas*, Addison-Wesley, Reading, Mass.

Chapter 3

The energy principle for MHD stability was given by I.B. Bernstein, E.A. Frieman, M.D. Kruskal and R.M. Kulsrud (1958) *Proc. Roy. Soc.* **A224**, 17.

A recent review of ideal MHD applied to fusion systems is given by J.P. Freidberg (1982) *Rev. Mod. Phys.* **54**, 801.

Clemmow and Dougherty (see above) give a thorough treatment of the connection between kinetic theory and MHD.

Chapters 4 and 5

The standard work on plasma waves is T.H. Stix (1962) *The Theory of Plasma Waves*, McGraw-Hill, New York.

A good account of waves in a cold plasma is given by T.J.M. Boyd and J.J. Sanderson (1969) *Plasma Dynamics*, Nelson, London, while a good account of the physical basis of Landau damping is given by Chen (see above).

Progress in radio frequency heating of plasmas may be monitored in the proceedings of the biennial *Symposia on Heating in Toroidal Plasma* (published by the Commission of the European Communities).

Chapter 6

Quasilinear theory was developed by W.E. Drummond and D. Pines (1964) *Ann. Phys. (N.Y.)* **28**, 478, and for a magnetized plasma by I.B. Bernstein and F. Engelmann (1966) *Phys. Fluids* **9**, 937.

Nonlinear wave interactions are discussed in R.C. Davidson (1972) *Methods in Nonlinear Plasma Physics*, Academic Press, London, and J. Weiland and H. Wilhelmsson (1977) *Coherent Non-linear Interaction of Waves in Plasmas*, Pergamon, Oxford.

A number of articles on parametric instabilities appear in A. Simon and W.B. Thomson (eds.) (1976) *Advances in Plasma Physics*, vol. 6, John Wiley & Sons, Chichester.

Chapter 7

Diagnostic methods are discussed in R.H. Huddlestone and S.L. Leonard (eds.) (1965) *Plasma Diagnostic Techniques*, Academic Press, London and H.R. Griem and R.H. Lovberg (eds.) (1970) *Methods in Experimental Physics*, vol. 9, Academic Press, London.

Chapter 8

The books by Chen and Myamoto and the collection of papers edited by Gill (see general references above), all provide further information on nuclear fusion research, particularly magnetic confinement.

Further information on inertial confinement can be found in H. Hora (1975) *Laser Plasmas and Nuclear Energy*, Plenum, London, and H. Motz (1979) *The Physics of Laser Fusion*, Academic Press, London.

Progress on all aspects of nuclear fusion may be followed in the *Proceedings of the Conferences on Plasma Physics and Controlled Fusion*, organized by the International Atomic Energy Agency.

A series of papers on the Earth's bow shock can be found in *Il Nuovo Cimento* (1979) **2C**, 653–859.

Applications of magnetohydrodynamics to solar physics are discussed in E.N. Parker, (1979) *Cosmical Magnetic Fields*, Clarendon Press, Oxford, and E.R. Priest (1982) *Solar Magnetohydrodynamics*, D. Reidel, Holland, while the dynamo problem is discussed in detail in H.K. Moffatt (1978) *Magnetic Field Generation in Electrically Conducting Fluids*, Cambridge University Press.

Pulsar magnetospheres are reviewed by F. Curtis Michel (1982) *Rev. Mod. Phys.* **54**, 1, and the acceleration of cosmic ray particles by L.O'C Drury (1983) *Rep. Prog. Phys.* **46**, 973.

Finally, the ideas outlined in section 8.10 on the use of plasma beat waves for particle acceleration are given by T. Tajima and J. M. Dawson (1979) *Phys. Rev. Lett.* **43**, 267 and T. Katsouleas and J.M. Dawson (1983) *Phys. Rev. Lett.* **51**, 392.

Index